QUANTITATIVE GEOGRAPHY

QUANTITATIVE GEOGRAPHY

Mohan Singh

ABD PUBLISHERS
Jaipur ■ New Delhi

ABD PUBLISHERS

Regd. Off. "Bony Residency", Gate No. 2
Opp. Tilak Public School, Vishweswariya Nagar,
Gopalpura Road, Jaipur - 302018, India
Ph: 0141 - 2761280, Telefax: 2761381
Email: abdpublishersjpr@gmail.com

Delhi Office: 102, 1st Floor, Satyam House,
4327/3, Ansari Road, Darya Ganj,
New Delhi - 110002, India
Ph: 011-45652440

---

visit us: www.oxfordbookcompany.com

Published in India, 2024

ISBN : 9788183763059

Price : ₹ 1995/-

Typeset by Shivangi Computer, Jaipur

Printed and bound in India by Thomson Press Ltd.,
New Delhi

Preface

Geostatistics deal with quantitative data analysis, specifically the application of statistical methodology to the exploration of geographic phenomena. Geostatistics is used extensively in a variety of fields including: hydrology, geology, petroleum exploration, weather analysis, urban planning, logistics, and epidemiology. The mathematical basis for geostatistics derives from cluster analysis, linear discriminant analysis and non-parametric statistical tests, and a variety of other subjects. Applications of geostatistics rely heavily on geographic information systems, particularly for the interpolation (estimate) of unmeasured points. Geographers are making notable contributions to the method of quantitative techniques. Geographic qualitative methods, or ethnographical; research techniques, are used by human geographers. In cultural geography there is a tradition of employing qualitative research techniques also used in anthropology and sociology. Participant observation and in-depth interviews provide human geographers with qualitative data.

Qualitative research is a method of inquiry appropriated in many different academic disciplines, traditionally in the social sciences, but also in market research and further contexts. Qualitative researchers aim to gather an in-depth understanding of human behaviour and the reasons that govern such behaviour. The qualitative method investigates the *why* and *how* of decision making, not just *what*, *where*, *when*. Hence, smaller but focused samples are more often needed, rather than large samples. Qualitative methods produce information only on the particular cases studied, and any more general conclusions are only hypotheses (informative guesses). Quantitative methods can be used to verify which of such hypotheses are truc. Pcrhaps the most traditional division in the way qualitative and quantitative research have been used in the social sciences is for qualitative methods to be used for exploratory (i.e., hypothesis-generating) purposes or explaining puzzling quantitative results, while quantitative methods are used to test hypotheses. This is because establishing content validity - do measures measure what a researcher thinks they measure? - is seen as one of the strengths of qualitative research. While quantitative methods are seen as providing more representative, reliable and precise measures through focused hypotheses, measurement tools and applied mathematics. By contrast, qualitative data is usually difficult to graph or display in mathematical terms.

Qualitative research is often used for policy and program evaluation research since it can answer certain important questions more efficiently and effectively than quantitative approaches. This is particularly the case for understanding how and why certain outcomes were achieved (not just what was achieved) but also answering important questions about relevance, unintended effects and impact of programs such as: Were expectations reasonable? Did processes operate as expected? Were key players able to carry out their duties? Were there any unintended effects of the program? Qualitative approaches have the advantage of allowing for more diversity in responses as well as the capacity to adapt to new developments or issues during the research process itself. While qualitative research can be expensive and time-consuming to conduct, many fields of research employ qualitative techniques that have been specifically developed to provide more succinct, cost-efficient and timely results. Rapid Rural Appraisal is one formalised example of these adaptations but there are many others. This book contains the fundamental and basic information of Quantities geography and the selection of contents makes it an appropriate book for the students.

Mohan Singh

Contents

1

Introduction

Geography is the study of the Earth and its lands, features, inhabitants, and phenomena. A literal translation would be "to describe or write about the Earth". The first person to use the word "geography" was Eratosthenes (276-194 B.C.). Four historical traditions in geographical research are the spatial analysis of natural and human phenomena (geography as a study of distribution), area studies (places and regions), study of man-land relationship, and research in earth sciences. Nonetheless, modern geography is an all-encompassing discipline that foremost seeks to understand the Earth and all of its human and natural complexities—not merely where objects are, but how they have changed and come to be. As "the bridge between the human and physical sciences," geography is divided into two main branches—human geography and physical geography.

Quantitative Geography is a lucid and comprehensive overview of the use of quantitative methods in spatial data analysis. It focuses on the philosophy informing spatial analysis and demonstrates the significant differences between modern quantitative methods and the methods associated with Geography's 'Quantitative Revolution' in the sixties. The text integrates a discussion of the application of quantitative methods with practical examples, and explains the philosophy of the new quantitative methodologies. Comprising a discussion of specific techniques, Quantitative Geography critically examines the profound difference in the use of those techniques since the quantitative revolution.

Quantitative Revolution

In the history of geography, the quantitative revolution was one of the four major turning-points of modern geography — the other three being regional geography, environmental determinism and critical geography). The quantitative revolution occurred during the 1950s and 1960s and marked a rapid change in the method behind geographical research. The main claim for the quantitative revolution is that it led to a shift from a descriptive (idiographic) geography to an empirical law making (nomothetic) geography. (Note: The quantitative revolution occurred earlier in Economics and Psychology and contemporarneously in Political science and other social sciences and to a lesser extent in History.)

Synopsis and Background

Many geography departments in the 1950s had recently separated from geology departments in the flux of postwar (World War II) enrolment. Because geologists of the time looked at geography as soft and unscientific, the feeling of many geographers was to persuade critics that geographers were not second-rate geologists. The changes during the 1950s through 1970s were not the introduction of mathematics into geography, but mathematics as a tool for explicit purposes and for statistical methodology and formal mathematical modelling.

In the early 1950s there was a growing sense that the existing paradigm for geographical research was not adequate in explaining how physical, economic, social, and political processes are spatially organized, ecologically related, or how outcomes generated by them are evidence for a given time and place. A more abstract, theoretical approach to geographical research has emerged, evolving the analytical method of inquiry.

The analytical method of inquiry led to the development of generalizations that are logically valid about the spatial aspects of a small set of closely defined events embodied in a wide range of natural and cultural settings. Generalizations may take the form of tested hypotheses, models, or theories and the research is judged on its scientific fit and its validity. Adoption of the analytical approach had helped geography become a more law-giving science, and the conception of the discipline as an idiographic field of study has become less acceptable starting in the 1980s.

The 1950s Crisis in Geography

During the late 1940s and early 1950s The crisis occurred for several reasons:

- The closing of many geography departments and courses in universities e.g. the abolition of the geography program at Harvard University in 1948
- Continuing division between Human and Physical geography-general talk of Human geography becoming an autonomous subject.
- Geography was seen (fairly or not) as overly descriptive and *unscientific*-there was, it was claimed, no explanation of *why* processes or phenomena occurred
- Geography was seen as exclusively educational-there were few if any applications of contemporary geography
- Continuing question of what geography is-Science, Art, Humanity or Social Science?
- After World War II technology became increasingly important in society and as a result nomothetic based sciences gained popularity and prominence

Debate raged predominantly (although not exclusively) in the U.S., where regional geography was the major philosophical school (European geography had never been uncomfortable with analytical methods).

All of these events presented a great threat to geography's position as an academic subject and thus geographers began seeking new methods to counter critique. Under the (somewhat misleading) banner of the scientific method, the quantitative revolution began.

The Revolution

The Quantitative Revolution began in the universities of Europe with the support of geographers and statisticians in both Europe and the United States. First emerging in the late 1950s and early 1960s, the Quantitative Revolution responded to the rising regional geography paradigm. Under the loosely defined banner of bringing 'scientific thinking' to geography, the quantitative revolution led to an increased use of computerized statistical techniques, in particular multivariate analysis, in geographical research. The newly adopted methods reflected an array of mathematical techniques that improved precision.

Some of the techniques that epitomize the quantitative revolution include:

- Descriptive statistics
- Inferential statistics
- Basic mathematical equations and models, such as gravity models
- Deterministic models, e.g. Von Thünen's and Weber's location models
- Stochastic models using concepts of probability.

Proponents of quantitative geography tended to present it as bringing science to geography. In fact, the particular contribution of the quantitative revolution was the huge faith placed in multivariate analysis and in particular methods associated with econometrics. It was also very strongly aligned with positive science and this would prove a major source of epistemological debate. The overwhelming focus on statistical modelling would, eventually, be the undoing of the quantitative revolution.

Many geographers became increasingly concerned that these techniques simply put a highly sophisticated technical gloss on an approach to study that was barren of theory.

Other critics argued that it removed the 'human dimension' from a discipline that always prided itself on studying the human and natural world alike. As the 1970s dawned, the quantitative revolution came under direct challenge.

Post-revolution Geography

The greatest impact of the quantitative revolution was not the revolution itself but the effects that came afterwards in a form of the spread of positivist (post-positivist) thinking and counter-positivist responses. The rising interest in the study of distance as a critical factor in understanding the spatial arrangement of phenomena during the revolution led to formulation of the first law of geography by Waldo Tobler. The development of spatial analysis in geography led to more applications in planning process and the further development of theoretical geography offered to geographical research a necessary theoretical background.

The greater use of computers in geography also led to many new developments in geomatics such as the creation and application of GIS and remote sensing. These new developments allowed geographers for the first time to assess complex models on a full-scale model and over space and time. The development of geomatics led to geography being reunited as the complexities of the human and natural environments could be assessed on new computable models. Further advances also led to a greater role of spatial statistics and modelling within geography. Eventually the quantitative revolution had its greatest impacts on the fields of physical, economic and urban geography.

The counter-positivist response from human geography was created in a form of behavioral, radical and humanistic geography.

The quantitative revolution also changed the structure of geography departments in the USA with many physical geographers being merged with geology departments or environmental science departments leaving the geography departments to become solely human geography oriented. Within the UK, there was a different response to the revolution with an increase of specialisation within the subject and ultimately the development of systematic geography with many subfields and branches.

Theories and Tendencies in Geography

In the case of geography the cultural turn can be understood in three ways and we will define it in terms of three types of cases:

1. Varying emphases in the research and study of cultural geographical subjects.

2. A new general trend affecting not only cultural geography but also geographical thinking as a whole and
3. A change in the overall intellectual attitude of geographers.

In this paper we shall limit ourselves to analyzing certain tendencies in geographical thinking that are a part of the cultural turn, highlighting-in the second case-what is happening today. This turn can be understood as described by Bergson in "L'Evolution Créatrice" (Bergson, 1907; Wright, 1947) when the attributed the idea of "becoming" to qualitative, evolutionary or extensive movement. The revaluation of what is qualitative would be a welcome alternative to the unbridled cult of the quantitative that has governed the last decades. Or, as Sorokin called it, to the "quantophrenia" to which geography has been no exception (Sorokin, 1964).

The evolutionary movement is to be noted when we contrast the normal interest that has always been manifested in the diverse processes of change on the face of the Earth with the accent now laid on the idea of ecological catastrophe. And finally the extensive movement might be considered as a more interactive and global vision of the Earth, going beyond the limits of traditional geography. Cultural tendencies are those that affect what is geographical *in toto,* in all its branches. These include geosophy, the purely sociological approach, an ideological approach and globalization-that we shall now proceed to analyze one by one.

Geosophy

Geosophy consists of an existentialist vision of geography, according to the definition of J.K.Wright (1945). It is a vision particularly centred not only on the perception of landscape, but also on other phenomena that occur on the surface of the earth It is, therefore, necessarily dominated by the psychological complexities of the relation of man to his environment.

Throughout a lengthy century, a geography impregnated with an excessively scientific approach led to a distrust of all that was not considered to be objective knowledge, discarding it as irrelevant. But, little by little, as a result of the indirect influence of psychology and a more intense knowledge of reality, more profound ideas came to be admitted as valid. Thus findings based on intuition or the threshold of perception began to be taken into consideration, largely as a result of the study of primitive cultures, which made it obvious that "civilized" man is less endowed with vital reflexes. We can also find in the chronicles of antiquity the attitude of the spontaneous observer; for example in Strabo, far from all rationalist influence, where spontaneous judgement is fully reflected. Or else in Ptolemy of whom Van Paasen writes: *"the naive, uncritical way in which, sporadically it is true, curiosities found in tales of travel are taken over"* (1957).

Democritus is possibly the major pre-Socratic philosopher of nature, (in connection with whom there is much we could say on this question of perception and knowledge). Louis Pauwels-and he is sometimes right-affirms that: *"his arguments were not those that we use* to-day, but they were both subtle and elegant, based on daily living. And his conclusions were *fundamentally correct"* (1980).

On the other hand, we had to reach the end of the XX century to recognize, as René Guenon (1969) has said, that there are types of knowledge that are not *"a question of erudition"* and *"that cannot be learned in any way from the reading of books"*.

Without pretending that the only valid knowledge is based on initiation, the esoteric, or is manifested as in the cave of Plato where the shadows are only visible thanks to the light, it is worth while remembering Saint Paul when in Corinthians XIII, 12, he contrasts what we may now know of God with what we shall find in Heaven "face to face": direct perception, intuitive knowledge.

What are we trying to say? Simply that there is another way of knowing objects than the scientific, experimental one. There exists a metaphysical knowledge, in the sense that Bergson understood it as a science (sic) of the real in itself, to which access is afforded by intuition. We are not in any way supporting here the occult, magic or purely imaginative theories that now invade geography from the field of philosophical pantheism, or that of "motherearth", leading the way to Teilhard de Chardin or Lovelock. As says Yves Galifret (1965): "*it* is necessary to distinguish clearly between reality and mystery. In the first case, evidence predominates and in the second, pure hypothesis. Imagination is not opposed to science except when it pretends to supplant it. The worst consists of converting it into an efficient end when, *at best, it can only be a means*".

A perception of the fantastic, of the extraordinary or of the infinitely subtle that is not recorded by any measuring apparatus and if it is recorded is not definitive, may nevertheless unexpectedly turn out to help an investigation. Such was the case with serendipity, where the unexpected and unsought after turned out to be the pathway to achievement. In all cases it is advisable to avoid confusing the two levels: that of methodical reasoning and that of instantaneous imagination. They both need to be tested to see if they are compatible. The fact that there are other technological alternatives apart from the leading *high tech* ones would also suggest that there exists an alternative type of scientific knowledge. A model personality for this view is Rudolf Steiner, who presumed that his effective discoveries in the field of biology (fertilizers that do not destroy the soil) and in medicine {the use of metals to modify metabolism) derived from theosophical or purely magic neo-pagan doctrines. All of this is particularly relevant to geography when it is a case of landscape, something that was undeniably rediscovered during the XX century with the pioneer work of Carl Sauer (1925).

Incidentally, a distinguished British town planner, Percy E.A. Johnson Marshall, once heard his colleagues scientifically arguing as to what should be the defining terms of reference for limiting a physical-planning region designed to protect the outskirts of Edinburgh. He then opted for no less than the unexpected definition of *eye sore* (what is damaging to vision- and also a pun on *I saw*!). With this phrase he distinguished the landscape worthy of preservation from what he considered to be irremediably lost. He was referring in the second case to what the Germans call *raublandschaft* (a landscape that has been ransacked or looted). Although the objective analytical process that takes into account data such as land eroded by wind or water, irrecoverable waste dumps, irrational deforestation, obsolete and antiaesthetic structures cannot be ignored, nevertheless, the *eye sore* method, eminently intuitive and sensitive as it is, has proved to be as accurate as it is ineffable-and provides a speedy short cut to the same conclusions as those reached by a rigorously rational method. It is also necessary to recognize that there are landscapes it is hard to read at first sight and whose systematic description does not fully define them. There are cases in which we should be forced to turn to the concept of *gestalt*, an expression that is untranslatable out of the psychological context, where, as is the case of human faces, a simple list of features would not lead to identification. Thus the technique of the *identikit* can only be applied by trial and error; all of which confirms the thesis that the whole-be it in the case of human faces or of the geographical *faces* that conform the landscape is more than the sum of the parts. It is perhaps superfluous to say there are multiple indirect factors that complement the visualization of landscape-such as narrative literature, legends, poems-or the chronicles of explorers, mountaineers and native inhabitants. All of these elements help to refine the *habitus* of contemplation, a searching inquiry into the mysterious (all that is not evident *de visu*) and the discovery that what appears before us is symbolic, ever containing hidden significance. But to accede to these paths a certain dose of *sympathy* is required-attraction to the connatural, as defined by Saint Thomas Aquinas, a dose of affectivity that facilitates the opening up of the visual reception process, without which it is impossible to end up by really knowing a landscape. As Saint Augustine once said of man, but is equally applicable, by analogy, to landscape: *nemo nisi per amicitiam cognoscitur* (it is necessary to establish friendship to understand him). All of which supports the idea of the importance of the point of view and the identity of the observer vis-à-vis the object.

Saint Augustine refers to another aspect that stresses the importance of the subjective factor in De Catequizante Rudibus when he speaks "...of what happens to us when we show visitors imposing cities and panoramas long familiar to us. Our pleasure is renewed with the novelty of theirs..." (Obras, 12,15). Seeing the same *thing through other eyes*, we discover nuances that we never saw the first time round or as a result of our customary routine. On the other hand, surrounded by the popular culture of the *homo videns* (Sartori, 1997). where we are swamped with images owing to our abusive use of television, our perception suffers a major interference a priori, as when a child *discovers* the elephant at the zoo long after he has read about him in a children's story book N. Whitehead (1932). And in this sense, the teaching of geography implies a two-edged weapon, for uncontrolled subjectivity can be a help but also a deviation from the real objective.

The Sociological Bias

In the world of today the social element eclipses the personal, the individual who used to be the main center of interest. That the collective view prevails in the analysis of all human activities is today beyond discussion. It is not surprising that this tendency has affected geography and has even led to an artificial dialectic between human and social geography. But there are other factors, also, that enable us to talk of the sociological derivation, when the social becomes the subject and the geographical the predicate. So we should not be too surprised that many research fellows become drawn to the social derivation path. They finally end up in a *reductionism* with all things centering on la *question sociale*-the class struggle or *la Bête sociale* as Simone Weil liked to call it-apart from their then proceeding to abandon all that is specifically geography.

A patent example of this is to be seen in the case of urban geography. The cities of to day are a seedbed of social problems owing to the increasing diversity of their functions and the density of their inhabitants. This is so to such an extent that even town planners-those who have the job of ordering urban space-become submerged in conflicts rent with an ideological connotation instead of proceeding to solve spatial issues in the light of common sense. In this way town planning loses in terms of its character as a professional job to be done and as a university training to become diluted into the so-called urban studies.

The first country to privilege the growing sociological approach to urbanism was the United States, where they converted into a hyperbole what was originally a valuable contribution: the urban ecology of the school of Chicago. Thus post-graduate degrees were granted to students who had graduated in economy, law, or social sciences, but had never been taught to read a plan or a map and lacked the spatial sense so indispensable to the exercise of the profession of town planner. Undoubtedly the borderline between auxiliary problems and geography *per se* is not always clear. For example, at a recent symposium on the geography of religions1 papers were presented that were presumed to belong to the field of geography because they alluded to such things as sanctuaries, (which obviously have a geographical location), the religions of immigrants (people coming from other lands), or the conflicts between imported sects-from other regions-and the religion of the country concerned. But in these cases, and many others that could also be invoked, the researchers usually follow the line of least resistance and end up submerged in social problems, easier and more obvious to deal with, rather than concentrating on the geographical content involved.

An exception to this have been, for example, the studies of Gaston Bardet on urban sociology that he placed under the heading of socio-topography, or the surveys of Father Lebret, strictly geographical in character and never distracted by ideological issues. More recently geography, largely owing to computer facilities-the GIS-has been seen to deal particularly with the purely formal aspects of distribution, rather that those of geographical content which concentrate on a deeper interaction between man and the environment. And this is equally true of the interest in processes of dissemination, as though the Earth were an abstract plane on which social phenomena can be verified.

Where the cultural turn in geography, in so far as it is dominated by sociology, is particularly evident is in the field of tuition. There it is gradually losing its character as an independent subject by becoming submerged in an area of social problems-while forfeiting its function as a bridge between the natural sciences and the humanities. "To integrate geography into wider areas of knowledge is an absurd idea, as though geography were not already, in itself, an area of knowledge", writes a professor (Pickenhayan). In 1992, the Commission on Education of the IGU found it necessary to formulate an "International Declaration on Education in Geography" in which special emphasis was laid on the fact that "*Geography should be considered a leading subject both in primary and secondary schools*".

Why should such a statement be necessary if geography was not being diluted by the social sciences, particularly under the aegis of the sociologists? Or otherwise taken over by the environmental problem, reaching a point where people talk about "environmental geography" with a markedly critical bias, rather than in a spirit of objective analysis, as though this were something new. The idea, in many cases, seems to be to try and stimulate the interest of the students. This might be acceptable as a teaching recourse if it did not lead to the study of a science based exclusively on questionnaires, inexorably concealing what is essential to the benefit of what is contingent.

Proof of the advance of the sociological influence is that statistics, based on surveys, prevail over maps in the teaching textbooks. This facilitates an a-spatial and even abstract conception, manifested in the ideological approach we have already mentioned and the globalization that we shall be dealing with later.

To summarize, this is the de-territorialization (devaluation of the concept of territory) that we have dealt with *in extenso* elsewhere; an idea that wins people over with the simplistic argument that "territory has a more vulnerable value than capital, work or know-how". And so, in the last resort, as we are now living in the times of the "global village", territory is no longer fashionable (Randle, 1999).

The Ideological Approach

Geography has always been a sounding box for fashionable philosophical trends, visible or invisible, Carthesian, mechanicist, determinist or materialist – as in the case of Marxist geography. The positivism of Reclus has left its mark and more recently, we would repeat, there is that of Bergson and then later comes Existentialism. In this last case the distinction between subject and object is downgraded and even the knowledge of reality appears to be subject to vital experiences.

To a lesser extent, owing to their more limited philosophical dimension, a correlation can be established between systems of thinking and the so-called "General Theory of Systems" of von Bertalanffy or with the structuralism of Levy Strauss. However, as Paul Claval (1998) has rightly said, these influences "are no longer fashionable, possibly because many of them have been superficial, merely a passing fad. And, finally, the theory of the catastrophe or those of chaos have been used to render banal simple causality. We also note a certain concomitance between analytical philosophy and neo-positivism – with their obviously mathematical basis-and location analysis or other types of quantitative geography.

However, it is one thing to say that ideas have always inspired geography (or rather, geographers) and another to detect the influence of ideologies, or rather speculations, unrelated to what is real, that soon become fossilized. For an ideology implies a commitment to a rigid formula and an exaggerated wish to impose the prevailing philosophy, preferring to persuade rather than to demonstrate.

The ideological discourse that expresses the beliefs and opinions of a given group constitutes, implicitly or explicitly, a call to action to try and impose these convictions; something that we have seen introduced into geography over the last decades. Radical geography is an extreme, an *altogether* example of the ideological approach, although a more subtle case is the so-called *green house effect*, now converted into an ideology. Not devoid of scientific arguments, it has been easily converted into a *doomsday* vision with the corresponding denunciation, and has not hesitated to invoke false arguments when required. This all began when some geographers alleged that geography could not be separated from the praxis, and hence it was inconceivable that it should not be involved in the definition of values. All previous geography was loosely described as positivist with a subliminal message to the effect that it had been an accomplice of social injustice. And thus the geography of conflicts was born with the risk (or deliberate intention) of exacerbating them and even of creating such conflicts where they did not already exist. That this was an ideological derivation rather than any true contribution to the treasures of geography is shown by the fact that no one would dream of reproaching nuclear physicists for not having incorporated into their scientific approach the consequences of the atomic bomb.

All of this is the tip of the rationalist ideological iceberg. And at the other extreme we find an irrational ideological approach. It is not perhaps too far-fetched to imagine that in the near future the cultural turn in geography will also come to incorporate an esoteric element, now currently so fashionable throughout the world.

Cultural geography provides a real hoard of information for research on this issue, granted that in ancient civilizations such as those of China all spatial references had a religious significance, especially where it was a case of orientation (Boyd, 1966). And in ancient Greece2, the site itself was viewed as sacred long before the construction of the temple. In China the Taoist religion preached the veneration of nature and its contemplation, creating a world of mysticism around the cardinal points. In ancient Greece a religious significance was attributed to the outline of the hills and the siting of the temples was conditioned by the interpretation of the landscape, dominated by the sense of the sacred. Each temple made use of the elements in the landscape, such as an enclosed valley that would serve as a natural *megaron,* as described by Scully, or a conical hill to be a cardinal point or Twin Peaks an axis. Mount Olympus, where no temple was required to be built for it to be considered a sanctuary, constitutes the clearest example of the relationship between religion and topography. Apart from this, all primitive cultures, including Rome, have privileged the notion of Nature as "Mother Earth" (*tellus mater*) attributing to it the role of fertility in all senses. Modernity, and with it the development of geography as a science, rejected these approaches centering all explanation on reason, but falling into an almost one-way rationalism that excluded any other consideration. And as a reaction to this view, the XX century witnessed the development of a tendency to discover the subjective factor that always modifies, if not entirely, our view of geographical reality.

This in turn has led to a tendency to render sacred things that are in no way numinous, but are simply separated from what is ordinary by being exceptional. Such is the case of a sporting ace, a diva, or an extraordinary site that may all end up by being conceded supernatural characteristics just because they are out of the ordinary.

We have already said it is not surprising that this tendency in contemporary culture should have an impact on new geographical trends. Were there not already suggestive articles along these lines in the magazine "Janus" in the '60s and before that in "Planete"? And although this may have been merely a passing phase, it is not unforeseeable that it may break out again since the ecological approach, based on so-called "scientific" hypotheses, is now reverting to new types of primitive animism. This is particularly clear in the work of James Lovelock (1979; 1988), with his hypothesis *Gaia* in which he considers the planet as an organic whole. His is a globalistic vision in physical terms that does not fail to have certain links with the politically globalist outlook, something we shall now look into it in more detail.

The search for an alternative approach to the purely scientific has led to the demonstration that what is rational does not cover all the field of geographical description, and this is undoubtedly to a certain extent understandable. Nevertheless, the fact that there are certain indefinable, ineffable characteristics in the landscape could lead to an attitude of reverent silence, and not necessarily to the prevailing temptation that falls into what Rudolf Otto (1925) describes as "extreme verbose mysticism"; what we would describe as a charlatan attitude. In Nature there are elements that are totally inaccessible to understanding in terms of concepts (as is true in a different field of what is beautiful). Which means that they cannot be exhaustively defined in terms of the traditional geographical descriptive process of a systematic character. A mountain may inspire an overwhelming reaction, a plain may suggest infinity, a distant mountain chain may awake our curiosity as to its hidden landscape. But these subjective impressions are in no way opposed to, and do not contradict, irreplaceable objective notions. Apart from their immanent significance, they awaken in us the feeling of the transcendental.

As is always the case with what is sacred, in generic terms it may lead us to a sacralization of the profane, to which we have already alluded. Something along similar lines occurs with values when they are maintained in purely formal terms and are not identified with truth, goodness and beauty.

A geography that is limited to the formal approach comes to classify places – always on a hypothetical basis – in terms of greater or lesser "energy" they may have, with the idea that this energy can be absorbed by human beings for the benefit of holistic health (physical and spiritual). A classical example of such a case is the transfer of a current hydrological concept (where energy is understood as the drainage capacity of the land) to the sphere of arbitrary subjectiveness, something based on fantasy with no scientific basis whatever, or-what is worse-actively opposed to it. With regard to the gnosis-eclectic doctrines that claim to initiate people into secrets hidden in Nature-this is gaining ground among the intellectuals *à la page* who disseminate a pantheistic vision of the world. Such a view is particularly exploited by the so-called *New Age* now creating a subculture (including a somewhat shaky spirituality). Why should this not also affect an irrational vision of the Earth?

The Earth, seen from this angle, has become just one more planet (the name of the magazine was not chosen by chance) thereby reducing the exclusive protagonism of man and of universal culture in order to dilute them in an interplanetary cosmos, a self-created universe that is a perfect entelechy. And with this view we find Carl Sagan with his cosmic egg or Frijof Capra with his neo-Taoism. It can be argued that none of this has really penetrated the geographical academic environment up to now, which is true. But it is worth while asking two questions: 1) Does it not all have a certain influence on the thinking that we have been discussing? And 2) Did not all of this begin with the Marxist school of geography? In both cases owing to the indifference of the majority.

Globalization

The global view of the Earth has been strengthened by its tie-in with economic, social and political trends that, strictly speaking, are not relevant, but have an influence on public opinion. Curiously enough, the opening-up implied by what is global in spatial terms has also been transferred to the time factor, involving other dimensions such as the geological and the biological, all of which leads to an evolutionary interpretation that is not necessarily geographical.

A consideration of the Earth as a living being, apart from its metaphorical nature and the argument as to the genesis of life on this planet with all the scientific interest involved, adds nothing at all to the idea of connectivity, inherent in the geographical view. Even the holistic vision of Nature as an inseparable unit provides nothing new for geography.

Are we not witnessing something similar to the concept of the equivalence of living bodies as postulated by Ratzel when he compared them with national states? And Vidal de la Blache, in agreement with Ritter, when emphasized the unity and interactivity of the earth -- with his concept of the *milieu* ? He did not, however, consider it necessary to deal with it undividedly, but rather the contrary "to study separately what nature has put together".

What is new would appear once again to consist of simply moving away from the principles of classical geography – often out of ignorance – supposing that the new approaches will make a substantial contribution. This is not, however, the case of the *Gaia* of Lovelock with the "unashamedly teleological idea that the Earth is a super-organism". Granted that this idea is very attractive and cannot be ignored, it should be noted that it is a question of a capricious interpolation of neovitalism, something quite foreign to geographical methodology.

Is it not strange that there should be geographers led astray by those who argue in favor of globalization based on reasons that are more economic or political than of any natural origin? Thus we are those who sustain the metaphor that "the world has grown smaller" or that distances have been cut down and territory is no longer of importance. And those who invoke the false dialectic between natural resources and technology, propagating the equally false idea that geography no longer has the relevance of times past.

We must be on guard against the temptation to devalue objective data, in the case of distances measuring them solely in terms of travel time or the cost of transport, substituting analogical concepts for the facts. Sensorial (and even extra-sensorial) perception may serve as a complement to a rational notion but they can never replace it.

The globalist mentality goes far beyond concrete global realities and often maintains exaggerated ideas as to the impact that globalization can have on geography. It is also dangerous to talk of post-modern geographies as though it were possible to establish limits to modern geography. There is, in fact, only a single geography, with a centuries old history in which it has been principally developed, along with the mainstream from which have sprung all varieties of contemporary geographical thinking. Basically the only post-phenomenon is the fashionable adoption of labels, which seems to reduce everything to a collection of diverse and incoherent tendencies.

Globalization, by postulating the theory of the progressive disappearance of territorial sovereignty, tends to undermine the credibility of geopolitics as it was conceived over most of the XX century. Apart from the fact that there may have been ideological factors leading to a loss of prestige for certain tendencies in geopolitics, it is now necessary for us to reconsider this whole subject. According to Huntington (1996), if not nations, then civilizations are going to clash in the near future, each seeking the domination of space.

Globalization is not a painless process. It involves huge migratory flows, demographic explosions in certain regions and a resistance to mixed blood that aggravates ethnic conflicts, all of which implies an inescapable re-thinking of geopolitics. There would seem to be a real plot against any attempt to revive geopolitics. "This ideology particularly distrusts geopolitics (a science that manipulates geographical realities) and considers that nations and religions are nothing more than visions of the spirit (Yves Lacoste would call them *representations*). The doorways of the French universities therefore remain closed today to geopolitics (Chauprade, 2002).

De-territorialization is the most typical effect of the introduction of the ideological approach to geography, a tendency that developed fast towards the end of the XX century. And whether owing to the growing influence of either Marxist or capitalist economic thinking, the effect was the same. In both cases the value of territory is questioned and, as a consequence, the very relevance of the geographic factor is placed at risk.

Geography is one of the oldest sciences that have formed their identities through fundamental research focused on developing new theories and methods, but also by solving specific spatial, social and economic problems. The awareness of importance of using geographic skills and spatial ways of thinking has greatly increased in the last decades. This is foremost a result of increasing challenges in the contemporary world, which can be largely attributed to scarcity and a rapid depletion of natural and social resources.

On the other hand, the evolution of geo-information techniques has offered a new approach in solving variety of global problems, including spatial management. At the same time, the last decades have seen an increased awareness of the importance of space, the so-called spatial turn in social sciences, which has provided an opportunity for the affirmation of geography in a new theoretical discourse of understanding the space and place. Thanks to its fundamental characteristic as a bridge between nature and society coupled with potential benefits from application of new information techniques, geography should definitely become one of the key sciences of the 21st century. However, applied geography today often takes place outside the academia, resulting in theory and practice becoming more separated, while geography tools and approaches are often used by professionals from other fields. At the same time the ideas of multidisciplinary approach in solving complex issues unfortunately, are sometimes far from the reality. This is exactly the reason why there is a need for an academic discussion about past experiences and future potentials of applied geography focused on problem-solving research in all geographic disciplines.

Geography was a relative latecomer as an established scientific discipline. As 19th century geographers struggled to align their subject with the positivistic emphasis of the times, they were apt to borrow both methods and metaphors from other sciences.

One such stolen concept was that of evolution – yet by applying it to nascent discipline of geopolitics they created a monster that would come to inform the ambitions of the Third Reich.

The Scientific Revolution and Geography

The exact nature of the scientific revolution, and indeed the question of whether it really occurred at all, is still debated by many historians. But it is generally considered to have begun during the 16th century, when great thinkers such as Copernicus, Bacon, Hohenheim and Kepler initiated a new paradigm of scientific investigation.

Geography, however, did not come to be seen as a discipline in its own right until part way through the 19th century. This transition was marked by increasing attempts to apply scientific theories to geographical entities – such as cities and nations. In devising new theories geographers tended to borrow heavily from the natural sciences.

Influences of Evolution on Geographical Thought

One such idea was the theory of evolution. In fact, two conflicting theories of evolution were both incorporated into early geopolitical works – namely, Lamarck's environmental determinism, outlined in the 1809 work *Philosophie Zoologique,* and the theory of natural selection proposed by Darwin's 1859 *On The Origin of Species*.

The two theories provided subtle but nevertheless conflicting metaphors for evolution. Environmental determinism posited a kind of natural order and inevitability; Darwin's natural selection lent a perceived scientific and moral backing to the domination of the weak by the strong. Both were incorporated into theories of the state-the former by Halford Mackinder and the latter by Friedrich Ratzel.

'Lebensraum' and the Heartland Theory

Ratzel was a German scientist with interests in a number of fields, including zoology and geography.

In 1901 he wrote of the natural development of nations, basing his observations on theories of evolution and historical examples from the formation of the British and French empires. He borrowed heavily from Darwin and argued that a nation was like an organism, coining the term *Lebensraum*-which in German means literally 'living space'-in relation to a nation's need for space to grow.

British geographer Mackinder, on the other hand, built more on environmental determinism. He described the history of the world in terms of geographical regions and the struggle of nations to control key regions. He divided the world into sections such as the Heartland and the Periphery and argued that nations in the heartland naturally assumed more power than those on the periphery. In Mackinder's geopolitics, control of the Heartland was key.

Geopolitik and the Rise of the Third Reich

During the 1920s, Karl Haushofer combined these ideas into a doctrine he termed *Geopolitik,* which was heavily used by the Third Reich both before and during the war.

Haushofer went further than Ratzel by identifying not only the concept of *Lebensraum* but also the natural right of a strong nation to expand into living space occupied by a weak nation. He also identified the land to the east of Germany – occupied by Poland – as a key strategic direction for that expansion due its location in the Heartland.

During the 1930s there was a general feeling of resentment among the German people, who felt humiliated after the First World War defeat and the Treaty of Versailles. In this environment the Third Reich was able to gain significant support by appealing to nationalistic fervour. Building on Haushofer's work, they championed Germany's natural superiority and moral right to expand, building a support base over the early part of the decade until Hitler was elected Chancellor.

The subsequent German invasion to the East was precipitated by Mackinder's Heartland theory, and during the war the Nazis published propaganda, such as *The War In Maps,* that sought to prove scientifically the inevitably of Germany's victory due to the natural order outlined in *Geopolitik.* Throughout the war, the natural right of the German people was championed by Haushofer's legacy.

Geopolitical Influences in History

Thus the 19th century struggle for legitimacy within geography contributed in many ways to the events of the Second World War. It is however far too simplistic to suggest that none of these events could have happened without the development of *Geopolitik.* There were many political and economic forces at work and the tensions that existed made it likely that conflict would have occurred in one way or another.

What can be said is that Haushofer's ideas significantly furthered the cause of the Nazis, and helped them convert the nationalistic fever that brought them to power into a moral crusade for a new German Empire. Without such an academic grounding, and without early geopolitical theories based on evolutionary ideas, it is possible that the Third Reich may never have come to power and World War II may not have happened in the way that it did.

2

Geography and Three Space Dimensions

Space is the boundless, three-dimensional extent in which objects and events occur and have relative position and direction. Physical space is often conceived in three linear dimensions, although modern physicists usually consider it, with time, to be part of the boundless four-dimensional continuum known as spacetime. In mathematics one examines 'spaces' with different numbers of dimensions and with different underlying structures. The concept of space is considered to be of fundamental importance to an understanding of the physical universe although disagreement continues between philosophers over whether it is itself an entity, a relationship between entities, or part of a conceptual framework.

Many of the philosophical questions arose in the 17th century, during the early development of classical mechanics. In Isaac Newton's view, space was absolute-in the sense that it existed permanently and independently of whether there were any matter in the space. Other natural philosophers, notably Gottfried Leibniz, thought instead that space was a collection of relations between objects, given by their distance and direction from one another. In the 18th century, Immanuel Kant described space and time as elements of a systematic framework that humans use to structure their experience.

In the 19th and 20th centuries mathematicians began to examine non-Euclidean geometries, in which space can be said to be *curved*, rather than *flat*. According to Albert Einstein's theory of general relativity, space around gravitational fields deviates from Euclidean space. Experimental tests of general relativity have confirmed that non-Euclidean space provides a better model for the shape of space.

Philosophy of Space

Leibniz and Newton

In the seventeenth century, the philosophy of space and time emerged as a central issue in epistemology and metaphysics. At its heart, Gottfried Leibniz, the German philosopher-mathematician, and Isaac Newton, the English physicist-mathematician, set out two opposing theories of what space is. Rather than being an entity that independently exists over and above other matter, Leibniz held that space is no more than the collection of spatial relations between objects in the world: "space is that which results from places taken together". Unoccupied regions are those that *could* have objects in them, and thus spatial relations with other places. For Leibniz, then, space was an idealised abstraction from the relations between individual entities or their possible locations and therefore could not be continuous but must be discrete. Space could be thought of in a similar way to the relations between family members. Although people in the family are related to one another, the relations do not exist independently of the people. Leibniz argued that space could not exist independently of objects in the world because that implies a difference between two universes exactly alike except for the location of the material world in each universe. But since there would be no observational way of telling these universes apart then, according to the identity of indiscernibles, there would be no real difference between them. According to the principle of sufficient reason, any theory of space that implied that there could be these two possible universes, must therefore be wrong.

Newton took space to be more than relations between material objects and based his position on observation and experimentation. For a relationist there can be no real difference between inertial motion, in which the object travels with constant velocity, and non-inertial motion, in which the velocity changes with time, since all spatial measurements are relative to other objects and their motions. But Newton argued that since non-inertial motion generates forces, it must be absolute. He used the example of water in a spinning bucket to demonstrate his argument. Water in a bucket is hung from a rope and set to spin, starts with a flat surface. After a while, as the bucket continues to spin, the surface of the water becomes concave. If the bucket's spinning is stopped then the surface of the water remains concave as it continues to spin. The concave surface is therefore apparently not the result of relative motion between the bucket and the water. Instead, Newton argued, it must be a result of non-inertial motion relative to space itself. For several centuries the bucket argument was decisive in showing that space must exist independently of matter.

Kant

In the eighteenth century the German philosopher Immanuel Kant developed a theory of knowledge in which knowledge about space can be both *a priori* and *synthetic*. According to Kant, knowledge about space is *synthetic*, in that statements about space are not simply true by virtue of the meaning of the words in the statement. In his work, Kant rejected the view that space must be either a substance or relation. Instead he came to the conclusion that space and time are not discovered by humans to be objective features of the world, but are part of an unavoidable systematic framework for organizing our experiences.

Non-Euclidean Geometry

Euclid's *Elements* contained five postulates that form the basis for Euclidean geometry. One of these, the parallel postulate has been the subject of debate among mathematicians for many centuries. It states that on any plane on which there is a straight line L_1 and a point P not on L_1, there is only one straight line L_2 on the plane that passes through the point P and is parallel to the straight line L_1. Until the 19th century, few doubted the truth of the postulate; instead debate centred over whether it was necessary as an axiom, or whether it was a theory that could be derived from the other axioms. Around

1830 though, the Hungarian János Bolyai and the Russian Nikolai Ivanovich Lobachevsky separately published treatises on a type of geometry that does not include the parallel postulate, called hyperbolic geometry.

In this geometry, an infinite number of parallel lines pass through the point *P*. Consequently the sum of angles in a triangle is less than 180° and the ratio of a circle's circumference to its diameter is greater than pi. In the 1850s, Bernhard Riemann developed an equivalent theory of elliptical geometry, in which no parallel lines pass through *P*. In this geometry, triangles have more than 180° and circles have a ratio of circumference-to-diameter that is less than pi.

Gauss and Poincaré

Although there was a prevailing Kantian consensus at the time, once non-Euclidean geometries had been formalised, some began to wonder whether or not physical space is curved. Carl Friedrich Gauss, the German mathematician, was the first to consider an empirical investigation of the geometrical structure of space. He thought of making a test of the sum of the angles of an enormous stellar triangle and there are reports he actually carried out a test, on a small scale, by triangulating mountain tops in Germany.

Henri Poincaré, a French mathematician and physicist of the late 19th century introduced an important insight in which he attempted to demonstrate the futility of any attempt to discover which geometry applies to space by experiment. He considered the predicament that would face scientists if they were confined to the surface of an imaginary large sphere with particular properties, known as a sphere-world.

In this world, the temperature is taken to vary in such a way that all objects expand and contract in similar proportions in different places on the sphere. With a suitable falloff in temperature, if the scientists try to use measuring rods to determine the sum of the angles in a triangle, they can be deceived into thinking that they inhabit a plane, rather than a spherical surface. In fact, the scientists cannot in principle determine whether they inhabit a plane or sphere and, Poincaré argued, the same is true for the debate over whether real space is Euclidean or not. For him, which geometry was used to describe space, was a matter of convention. Since Euclidean geometry is simpler than non-Euclidean geometry, he assumed the former would always be used to describe the 'true' geometry of the world.

Einstein

In 1905, Albert Einstein published a paper on a special theory of relativity, in which he proposed that space and time be combined into a single construct known as *spacetime*. In this theory, the speed of light in a vacuum is the same for all observers—which has the result that two events that appear simultaneous to one particular observer will not be simultaneous to another observer if the observers are moving with respect to one another. Moreover, an observer will measure a moving clock to tick more slowly than one that is stationary with respect to them; and objects are measured to be shortened in the direction that they are moving with respect to the observer.

Over the following ten years Einstein worked on a general theory of relativity, which is a theory of how gravity interacts with spacetime. Instead of viewing gravity as a force field acting in spacetime, Einstein suggested that it modifies the geometric structure of spacetime itself. According to the general theory, time goes more slowly at places with lower gravitational potentials and rays of light bend in the presence of a gravitational field. Scientists have studied the behaviour of binary pulsars, confirming the predictions of Einstein's theories and Non-Euclidean geometry is usually used to describe spacetime.

Mathematics

In modern mathematics spaces are defined as sets with some added structure. They are frequently described as different types of manifolds, which are spaces that locally approximate to Euclidean space, and where the properties are defined largely on local connectedness of points that lie on the manifold. There are however, many diverse mathematical objects that are called spaces. For example, function spaces in general have no close relation to Euclidean space.

Physics

Classical Mechanics

Space is one of the few fundamental quantities in physics, meaning that it cannot be defined via other quantities because nothing more fundamental is known at the present. On the other hand, it can be related to other fundamental quantities. Thus, similar to other fundamental quantities (like time and mass), space can be explored via measurement and experiment.

Astronomy

Astronomy is the science involved with the observation, explanation and measuring of objects in outer space.

Relativity

Before Einstein's work on relativistic physics, time and space were viewed as independent dimensions. Einstein's discoveries showed that due to relativity of motion our space and time can be mathematically combined into one object — spacetime. It turns out that distances in space or in time separately are not invariant with respect to Lorentz coordinate transformations, but distances in Minkowski space-time along space-time intervals are—which justifies the name.

In addition, time and space dimensions should not be viewed as exactly equivalent in Minkowski space-time. One can freely move in space but not in time. Thus, time and space coordinates are treated differently both in special relativity (where time is sometimes considered an imaginary coordinate) and in general relativity (where different signs are assigned to time and space components of spacetime metric).

Furthermore, in Einstein's general theory of relativity, it is postulated that space-time is geometrically distorted-*curved*-near to gravitationally significant masses.

Experiments are ongoing to attempt to directly measure gravitational waves. This is essentially solutions to the equations of general relativity, which describe moving ripples of spacetime. Indirect evidence for this has been found in the motions of the Hulse-Taylor binary system.

Cosmology

Relativity theory leads to the cosmological question of what shape the universe is, and where space came from. It appears that space was created in the Big Bang and has been expanding ever since. The overall shape of space is not known, but space is known to be expanding very rapidly due to the Cosmic Inflation. Alan Guth whom is known for his Inflationary theory, presented the first ideas in a seminar at Stanford Linear Accelerator Center on January 23, 1980.

Spatial Measurement

The measurement of *physical space* has long been important. Although earlier societies had developed measuring systems, the International System of Units, (SI), is now the most common system of units used in the measuring of space, and is almost universally used within science.

Currently, the standard space interval, called a standard meter or simply meter, is defined as the distance traveled by light in a vacuum during a time interval of exactly 1/299,792,458 of a second. This definition coupled with present definition of the second is based on the special theory of relativity in which the speed of light plays the role of a fundamental constant of nature.

Geography

Geography is the branch of science concerned with identifying and describing the Earth, utilizing spatial awareness to try and understand why things exist in specific locations. Cartography is the mapping of spaces to allow better navigation, for visualization purposes and to act as a locational device. Geostatistics apply statistical concepts to collected spatial data to create an estimate for unobserved phenomena.

Geographical space is often considered as land, and can have a relation to ownership usage (in which space is seen as property or territory). While some cultures assert the rights of the individual in terms of ownership, other cultures will identify with a communal approach to land ownership, while still other cultures such as Australian Aboriginals, rather than asserting ownership rights to land, invert the relationship and consider that they are in fact owned by the land. Spatial planning is a method of regulating the use of space at land-level, with decisions made at regional, national and international levels. Space can also impact on human and cultural behaviour, being an important factor in architecture, where it will impact on the design of buildings and structures, and on farming.

Ownership of space is not restricted to land. Ownership of airspace and of waters is decided internationally. Other forms of ownership have been recently asserted to other spaces — for example to the radio bands of the electromagnetic spectrum or to cyberspace.

Public space is a term used to define areas of land as collectively owned by the community, and managed in their name by delegated bodies; such spaces are open to all. While private property is the land culturally owned by an individual or company, for their own use and pleasure.

Abstract space is a term used in geography to refer to a hypothetical space characterized by complete homogeneity. When modelling activity or behaviour, it is a conceptual tool used to limit extraneous variables such as terrain.

Geomatics

Geomatics is the discipline of gathering, storing, processing, and delivering geographic information, or spatially referenced information.

Overview

Geomatics is fairly new, the term was apparently coined by B. Dubuisson in 1969 from the combination of geodesy and geoinformatics terms. It includes the tools and techniques used in land surveying, remote sensing, cartography, Geographic Information Systems (GIS), Global Navigation Satellite Systems (GPS, GLONASS, GALILEO, COMPASS), photogrammetry, and related forms of earth mapping. Originally used in Canada, because it is similar in French and English, the term geomatics has been adopted by the International Organization for Standardization, the Royal Institution of Chartered Surveyors, and many other international authorities, although some (especially in the United States) have shown a preference for the term *geospatial technology*.

A good definition can be found on the University of Calgary's web page titled "What is Geomatic Engineering?":

> *"Geomatics engineering is a modern discipline, which integrates acquisition, modelling, analysis, and management of spatially referenced data, i.e. data identified according to their locations. Based on the scientific framework of geodesy, it uses terrestrial, marine, airborne, and satellite-based sensors to acquire spatial and other data. It includes the process of transforming spatially referenced data from different sources into common information systems with well-defined accuracy characteristics."*

Similarly the new related field hydrogeomatics covers the geomatics area associated with surveying work carried out on, above or below the surface of the sea or other areas of water. The older term of hydrographics was too specific to the preparation of marine charts and failed to include the broader concept of positioning or measurements in all marine environments.

A geospatial network is a network of collaborating resources for sharing and coordinating geographical data, and data tied to geographical references. One example of such a network is the GIS Consortium's effort to provide "ready global access to geographic information" in a framework named the Open Geospatial Network.

A number of university departments which were once titled surveying, survey engineering or topographic science have re-titled themselves as geomatics or geomatic engineering. An example of this is the Department of Civil, Environmental and Geomatic Engineering at University College London.

The rapid progress, and increased visibility, of geomatics since 1990s has been made possible by advances in computer technology, computer science, and software engineering, as well as airborne and space observation remote sensing technologies.

3

Statistics in Geography

Statistics is the formal science of making effective use of numerical data relating to groups of individuals or experiments. It deals with all aspects of this, including not only the collection, analysis and interpretation of such data, but also the planning of the collection of data, in terms of the design of surveys and experiments.

A statistician is someone who is particularly well versed in the ways of thinking necessary for the successful application of statistical analysis. Often such people have gained this experience after starting work in any of a number of fields. There is also a discipline called *mathematical statistics*, which is concerned with the theoretical basis of the subject.

The word *statistics* can either be singular or plural. When it refers to the discipline, "statistics" is singular, as in "Statistics is an art." When it refers to quantities (such as mean and median) calculated from a set of data, *statistics* is plural, as in "These statistics are misleading."

Scope

Statistics is considered by some to be a mathematical science pertaining to the collection, analysis, interpretation or explanation, and presentation of data, while others consider it to be a branch of mathematics concerned with collecting and interpreting data. Because of its empirical roots and its focus on applications, statistics is usually considered to be a distinct mathematical science rather than a branch of mathematics.

Statisticians improve the quality of data with the design of experiments and survey sampling. Statistics also provides tools for prediction and forecasting using data and statistical models. Statistics is applicable to a wide variety of academic disciplines, including natural and social sciences, government, and business.

Statistical methods can be used to summarize or describe a collection of data; this is called *descriptive statistics*. This is useful in research, when communicating the results of experiments. In addition, patterns in the data may be modeled in a way that accounts for randomness and uncertainty in the observations, and are then used to draw inferences about the process or population being studied; this is called *inferential statistics*.

Inference is a vital element of scientific advance, since it provides a prediction (based in data) for where a theory logically leads. To further prove the guiding theory, these predictions are tested as well, as part of the scientific method. If the inference holds true, then the descriptive statistics of the new data increase the soundness of that hypothesis. Descriptive statistics and inferential statistics (a.k.a., predictive statistics) together comprise *applied statistics*.

History

Some scholars pinpoint the origin of statistics to 1663, with the publication of *Natural and Political Observations upon the Bills of Mortality* by John Graunt. Early applications of statistical thinking revolved around the needs of states to base policy on demographic and economic data, hence its *stat-* etymology. The scope of the discipline of statistics broadened in the early 19th century to include the collection and analysis of data in general. Today, statistics is widely employed in government, business, and the natural and social sciences.

Its mathematical foundations were laid in the 17th century with the development of probability theory by Blaise Pascal and Pierre de Fermat. Probability theory arose from the study of games of chance. The method of least squares was first described by Carl Friedrich Gauss around 1794. The use of modern computers has expedited large-scale statistical computation, and has also made possible new methods that are impractical to perform manually.

Overview

In applying statistics to a scientific, industrial, or societal problem, it is necessary to begin with a population or process to be studied. Populations can be diverse topics such as "all persons living in a country" or "every atom composing a crystal". A population can also be composed of observations of a process at various times, with the data from each observation serving as a different member of the overall group. Data collected about this kind of "population" constitutes what is called a time series.

For practical reasons, a chosen subset of the population called a sample is studied — as opposed to compiling data about the entire group (an operation called census). Once a sample that is representative of the population is determined, data is collected for the sample members in an observational or experimental setting. This data can then be subjected to statistical analysis, serving two related purposes: description and inference.

- Descriptive statistics summarize the population data by describing what was observed in the sample numerically or graphically. Numerical descriptors include mean and standard deviation for continuous data types (like heights or weights), while frequency and percentage are more useful in terms of describing categorical data (like race).
- Inferential statistics uses patterns in the sample data to draw inferences about the population represented, accounting for randomness. These inferences may take the form of: answering yes/no questions about the data (hypothesis testing), estimating numerical characteristics of the data (estimation), describing associations within the data (correlation), modelling relationships within the data (regression), extrapolation, interpolation, or other modelling techniques like ANOVA, time series, and data mining.

The concept of correlation is particularly noteworthy for the potential confusion it can cause. Statistical analysis of a data set often reveals that two variables (properties) of the population under consideration tend to vary together, as if they were connected. For example, a study of annual income that also looks at age of death might find that poor people tend to have shorter lives than affluent people. The two variables are said to be correlated; however, they may or may not be the cause of one another. The correlation phenomena could be caused by a third, previously unconsidered phenomenon, called a lurking variable or confounding variable. For this reason, there is no way to immediately infer the existence of a causal relationship between the two variables.

For a sample to be used as a guide to an entire population, it is important that it is truly a representative of that overall population. Representative sampling assures that the inferences and conclusions can be safely extended from the sample to the population as a whole. A major problem lies in determining the extent to which the sample chosen is actually representative. Statistics offers methods to estimate and correct for any random trending within the sample and data collection procedures. There are also methods for designing experiments that can lessen these issues at the outset of a study, strengthening its capability to discern truths about the population. Statisticians describe stronger methods as more "robust".

Randomness is studied using the mathematical discipline of probability theory. Probability is used in "Mathematical statistics" (alternatively, "statistical theory") to study the sampling distributions of sample statistics and, more generally, the properties of statistical procedures. The use of any statistical method is valid when the system or population under consideration satisfies the assumptions of the method.

Misuse of statistics can produce subtle, but serious errors in description and interpretation — subtle in the sense that even experienced professionals make such errors, and serious in the sense that they can lead to devastating decision errors. For instance, social policy, medical practice, and the reliability of structures like bridges all rely on the proper use of statistics. Even when statistics are correctly applied, the results can be difficult to interpret for those lacking expertise. The statistical significance of a trend in the data — which measures the extent to which a trend could be caused by random variation in the sample — may or may not agree with an intuitive sense of its significance. The set of basic statistical skills (and skepticism) that people need to deal with information in their everyday lives properly is referred to as statistical literacy.

Statistical Methods

Experimental and Observational Studies

A common goal for a statistical research project is to investigate causality, and in particular to draw a conclusion on the effect of changes in the values of predictors or independent variables on dependent variables or response. There are two major types of causal statistical studies: experimental studies and observational studies. In both types of studies, the effect of differences of an independent variable (or variables) on the behaviour of the dependent variable are observed. The difference between the two types lies in how the study is actually conducted. Each can be very effective. An experimental study involves taking measurements of the system under study, manipulating the system, and then taking additional measurements using the same procedure to determine if the manipulation has modified the values of the measurements. In contrast, an observational study does not involve experimental manipulation. Instead, data are gathered and correlations between predictors and response are investigated.

Experiments

The basic steps of a statistical experiment are:

1. Planning the research, including finding the number of replicates of the study, using the following information: preliminary estimates regarding the size of treatment effects, alternative hypotheses, and the estimated experimental variability. Consideration of the selection of experimental subjects and the ethics of research is necessary. Statisticians recommend that experiments compare (at least) one new treatment with a standard

treatment or control, to allow an unbiased estimate of the difference in treatment effects.

2. Design of experiments, using blocking to reduce the influence of confounding variables, and randomized assignment of treatments to subjects to allow unbiased estimates of treatment effects and experimental error. At this stage, the experimenters and statisticians write the *experimental protocol* that shall guide the performance of the experiment and that specifies the *primary analysis* of the experimental data.
3. Performing the experiment following the experimental protocol and analyzing the data following the experimental protocol.
4. Further examining the data set in secondary analyses, to suggest new hypotheses for future study.
5. Documenting and presenting the results of the study.

Experiments on human behaviour have special concerns. The famous Hawthorne study examined changes to the working environment at the Hawthorne plant of the Western Electric Company. The researchers were interested in determining whether increased illumination would increase the productivity of the assembly line workers. The researchers first measured the productivity in the plant, then modified the illumination in an area of the plant and checked if the changes in illumination affected productivity. It turned out that productivity indeed improved (under the experimental conditions). However, the study is heavily criticized today for errors in experimental procedures, specifically for the lack of a control group and blindness. The Hawthorne effect refers to finding that an outcome (in this case, worker productivity) changed due to observation itself. Those in the Hawthorne study became more productive not because the lighting was changed but because they were being observed.

Observational Study

An example of an observational study is one that explores the correlation between smoking and lung cancer. This type of study typically uses a survey to collect observations about the area of interest and then performs statistical analysis. In this case, the researchers would collect observations of both smokers and non-smokers, perhaps through a case-control study, and then look for the number of cases of lung cancer in each group.

Levels of Measurement

There are four main levels of measurement used in statistics:

- nominal,
- ordinal,
- interval, and
- ratio.

They have different degrees of usefulness in statistical research. Ratio measurements have both a meaningful zero value and the distances between different measurements defined; they provide the greatest flexibility in statistical methods that can be used for analyzing the data. Interval measurements have meaningful distances between measurements defined, but the zero value is arbitrary (as in the case with longitude and temperature measurements in Celsius or Fahrenheit). Ordinal measurements have imprecise differences between consecutive values, but have a meaningful order to those values. Nominal measurements have no meaningful rank order among values.

Because variables conforming only to nominal or ordinal measurements cannot be reasonably measured numerically, sometimes they are grouped together as categorical variables, whereas ratio and interval measurements are grouped together as quantitative or continuous variables due to their numerical nature.

Key Terms used in Statistics

Null Hypothesis

Interpretation of statistical information can often involve the development of a null hypothesis in that the assumption is that whatever is proposed as a cause has no effect on the variable being measured. The best illustration for a novice is the predicament encountered by a jury trial. The null hypothesis,

H0, asserts that the defendant is innocent, whereas the alternative hypothesis, H1, asserts that the defendant is guilty.

The indictment comes because of suspicion of the guilt. The H0 (status quo) stands in opposition to H1 and is maintained unless H1 is supported by evidence "beyond a reasonable doubt". However, "failure to reject H0" in this case does not imply innocence, but merely that the evidence was insufficient to convict. So the jury does not necessarily *accept* H0 but *fails to reject* H0.

Error

Working from a null hypothesis two basic forms of error are recognised:

- Type I errors where the null hypothesis is falsely rejected giving a "false positive".
- Type II errors where the null hypothesis fails to be rejected and an actual difference between populations is missed.

Error also refers to the extent to which individual observations in a sample differ from a central value, such as the sample or population mean. Many statistical methods seek to minimize the mean-squared error, and these are called "methods of least squares."

Measurement processes that generate statistical data are also subject to error. Many of these errors are classified as random (noise) or systematic (bias), but other important types of errors (e.g., blunder, such as when an analyst reports incorrect units) can also be important.

Confidence Intervals

Most studies will only sample part of a population and then the result is used to interpret the null hypothesis in the context of the whole population. Any estimates obtained from the sample only approximate the population value. Confidence intervals allow statisticians to express how closely the sample estimate matches the true value in the whole population. Often they are expressed as 95% confidence intervals. Formally, a 95% confidence interval of a procedure is any range such that the interval covers the true population value 95% of the time given repeated sampling under the same conditions. If these intervals span a value (such as zero) where the null hypothesis would be confirmed then this can indicate that any observed value has been seen by chance. For example a drug that gives a mean increase in heart rate of 2 beats per minute but has 95% confidence intervals of-5 to 9 for its increase may well have no effect whatsoever.

The 95% confidence interval is often misinterpreted as the probability that the true value lies between the upper and lower limits given the observed sample. However this quantity is more a credible interval available only from Bayesian statistics.

Significance

Statistics rarely give a simple Yes/No type answer to the question asked of them. Interpretation often comes down to the level of statistical significance applied to the numbers and often refer to the probability of a value accurately rejecting the null hypothesis (sometimes referred to as the p-value).

When interpreting an academic paper reference to the significance of a result when referring to the statistical significance does not necessarily mean that the overall result means anything in real world terms. (For example in a large study of a drug it may be shown that the drug has a statistically significant but very small beneficial effect such that the drug will be unlikely to help anyone given it in a noticeable way.)

Geographic Data

The geographic data that describes our world allows for city planning, flood prediction and relief, emergency service routing, environmental assessments, wind pattern monitoring and many other applications.

Geographic data is processed with Geographic information system (GIS) software which can, as one aspect of its functioning, produce maps.

In the United States, geographic data collected by central government is made available free of copyright for no more than the cost of distribution. The United States Census Bureau's TIGER Mapsurfer provides a web service and also offers data free for download.

TIGER allows you to build a geocoding facility with which to spatially locate addresses. Given the ability to geocode street addresses and other features, one can create a lot of interesting spatial analysis, location-based service, political campaigning apps and localised search services.

In the EU there is a European Union directive (INSPIRE directive) to establish shared standards between the different countries, accompanied by web viewing of rendered map data, and an unspecified license framework for geographic data.

Geocoding

Geocoding is the process of finding associated geographic coordinates (often expressed as latitude and longitude) from other geographic data, such as street addresses, or zip codes (postal codes). With geographic coordinates the features can be mapped and entered into Geographic Information Systems, or the coordinates can be embedded into media such as digital photographs via geotagging.

Reverse geocoding is the opposite: finding an associated textual location such as a street address, from geographic coordinates.

A geocoder is a piece of software or a (web) service that helps in this process.

Address Interpolation

A simple method of geocoding is address interpolation. This method makes use of data from a street geographic information system where the street network is already mapped within the geographic coordinate space.

Each street segment is attributed with address ranges (e.g. house numbers from one segment to the next). Geocoding takes an address, matches it to a street and specific segment (such as a block, in towns that use the "block" convention). Geocoding then interpolates the position of the address, within the range along the segment.

Example

Let's say that this segment (for instance, a block) of Evergreen Terrace runs from 700 to 799. Even-numbered addresses would fall on one side (e.g. west side) of Evergreen Terrace, with odd-numbered addresses on the other side (e.g. east side). 742 Evergreen Terrace would (probably) be located slightly less than halfway up the block, on the west side of the street. A point would be mapped at that location along the street, perhaps offset some distance to the west of the street centerline.

Complicating Factors

However, this process is not always as straightforward as in this example.

Difficulties arise when

- distinguishing between ambiguous addresses such as 742 Evergreen Terrace and 742 W Evergreen Terrace.
- attempting to geocode new addresses for a street that is not yet added to the geographic information system database.

While there might be 742 Evergreen Terrace in Springfield, there might also be a 742 Evergreen Terrace in Shelbyville. Asking for the city name (and state, province, country, etc. as needed) can solve this problem. Some situations require use of postal codes or district name for disambiguation. For example, there are multiple 100 Washington Streets in Boston, Massachusetts because several cities have been annexed without changing street names.

Finally, several caveats on using interpolation:

- The typical attribution of a street segment assumes that all "even" numbered parcels are on one side of the segment, and all "odd" numbered parcels are on the other. This is often not true in real life.
- Interpolation assumes that the given parcels are evenly distributed along the length of the segment. This is almost never true in real life; it is not uncommon for a geocoded address to be off by several thousand feet.

- Segment Information (esp. from sources such as TIGER) includes a maximum upper bound for addresses and is interpolated as though the full address range is used. For example, a segment (block) might have a listed range of 100-199, but the last address at the end of the block is 110. In this case, address 110 would be geocoded to 10% of the distance down the segment rather than near the end.
- Most interpolation implementations will produce a point as their resulting "address" location. In reality, the physical address is distributed along the length of the segment, i.e. consider geocoding the address of a shopping mall-the physical lot may run quite some distance along the street segment (or could be thought of as a two-dimensional space-filling polygon which may front on several different streets-or worse, for cities with multi-level streets, a three-dimensional shape that meets different streets at several different levels) but the interpolation treats it as a singularity.

A very common error is to believe the accuracy ratings of a given map's geocodable attributes. Such "accuracy" currently touted by most vendors has no bearing on an address being attributed to the correct segment, being attributed to the correct "side" of the segment, nor resulting in an accurate position along that correct segment. With the geocoding process used for U.S. Census TIGER datasets, 5-7.5% of the addresses may be allocated to a different census tract, while 50% of the geocoded points might be located to a different property parcel.

Because of this, it is quite important to avoid using interpolated results except for non-critical applications, such as pizza delivery. Interpolated geocoding is usually not appropriate for making authoritative decisions, for example if life safety will be impacted by that decision. Emergency services, for example, do not make an authoritative decision based on their interpolations; an ambulance or fire truck will always be dispatched regardless of what the map says.

Other Techniques

Other means of geocoding might include locating a point

at the centroid (center) of a land parcel, if parcel (property) data is available in the geographic information system database. In rural areas or other places lacking high quality street network data and addressing, GPS is useful for mapping a location. For traffic accidents, geocoding to a street intersection or midpoint along a street centerline is a suitable technique. Most highways in developed countries have mile markers to aid in emergency response, maintenance, and navigation. It is also possible to use a combination of these geocoding techniques-using a particular technique for certain cases and situations and other techniques for other cases.

Uses

Geocoded locations are useful in many GIS analysis and cartography tasks.

Geocoding is common on the web, for services like finding driving directions to or from some address, or finding a list of the geographically nearest store or service locations.

Geocoding is one of several methods of obtaining geographic coordinates for geotagging media, such as photographs or RSS items.

Privacy Concerns

The proliferation and ease of access to geocoding (and reverse-geocoding) services raises privacy concerns. For example, in mapping crime incidents, law enforcement agencies aim to balance the privacy rights of victims and offenders, with the public's right to know. Law enforcement agencies have experimented with alternative geocoding techniques that allow them to mask some of the locational detail (e.g., address specifics that would lead to identifying a victim or offender). As well, in providing online crime mapping to the public, they also place disclaimers regarding the locational accuracy of points on the map, acknowledging these location masking techniques, and impose terms of use for the information.

Reverse Geocoding

Reverse geocoding is the process of back (reverse) coding of a point location (latitude, longitude) to a readable address or place name. This permits the identification of nearby street addresses, places, and/or areal subdivisions such as neighborhoods, county, state, or country. Combined with geocoding and routing services, reverse geocoding is a critical component of mobile location-based services and Enhanced 911 to convert a coordinate obtained by GPS to a readable street address which is easier to understand by the end user.

Reverse geocoding can be carried out systematically by services which process a coordinate similarly to the geocoding process. For example, when a GPS coordinate is entered the street address is interpolated from a range assigned to the road segment in a reference dataset that the point is nearest to. If the user provides a coordinate near the midpoint of a segment that starts with address 1 and ends with 100, the returned street address will be somewhere near 50. This approach to reverse geocoding does not return actual addresses, only estimates of what should be there based on the predetermined range. Alternatively, coordinates for reverse geocoding can also be selected on an interactive map, or extracted from static maps by georeferencing them in a GIS with predefined spatial layers to determine the coordinates of a displayed point. Many of the same limitations of geocoding are similar with reverse geocoding. The accuracy and timeliness of the reference layer used to reverse geocode a coordinate will have a significant impact on the accuracy of the results.

Reverse geocoding services have typically not been public due to the need for extensive computing resources and currently updated and large databases. However, public reverse geocoding services are becoming increasingly available through APIs and other web services as well as mobile phone applications. These services require manual input of a coordinate, capture from a GPS, or selection of a point on an interactive map; to look up a street address or neighboring places. Examples of these services include the GeoNames reverse geocoding web service which has tools to identify nearest street address, place names, Wikipedia articles, country, county subdivisions, neighborhoods, and other location data from a coordinate. GeoNames uses the United States Census Bureau's tiger line data set as the reference layer for reverse geocoding. Google has also published a reverse geocoding API which can be adapted for online reverse geocoding tools, which uses the same street reference layer as Google maps.

Privacy Concerns

Geocoding and reverse geocoding have raised potential privacy concerns, especially regarding the ability to reverse engineer street addresses from published static maps. By digitizing published maps it is possible to georeference them by overlaying with other spatial layers and then extract point locations which can be used to identify individuals or reverse geocoded to obtain a street address of the individual. This has potential implications to determine locations for patients and/ or study participants from maps published in medical literature as well as potentially sensitive information published in other journalistic sources.

The ability to reverse engineer maps and obtain readable location information from static newspaper maps and hypothetical patient address maps has been examined. In one study a map of Hurricane Katrina mortality locations published in a Baton Rouge, Louisiana paper was examined. Using GPS locations obtained from houses where fatalities occurred, the authors were able to determine the relative error between the true house locations and the location determined by georeferencing the published map. The authors found that approximately 45% of the points extracted from the georeferenced map were within 10 meters of a households GPS obtained point. Another study found similar results examining hypothetical low and high resolution patient address maps similar to what might be found published in medical journals. They found approximately 26% of points obtained from a low resolution map and 79% from a high resolution map were matched precisely with the true location.

The findings from these studies raise concerns regarding the potential use of georeferencing and reverse geocoding of published maps to elucidate sensitive or private information on mapped individuals. Guidelines for the display and publication of potentially sensitive information are inconsistently applied and no uniform procedure has been identified. The use of blurring algorithms which shift the location of mapped points have been proposed as a solution. In addition, where direct reference to the geography of the area mapped is not required it may be possible to use abstract space on which to display spatial patterns.

Spatial Analysis

In statistics, spatial analysis or spatial statistics includes any of the formal techniques which study entities using their topological, geometric, or geographic properties. The phrase properly refers to a variety of techniques, many still in their early development, using different analytic approaches and applied in fields as diverse as astronomy, with its studies of the placement of galaxies in the cosmos, to chip fabrication engineering, with its use of 'place and route' algorithms to build complex wiring structures. The phrase is often used in a more restricted sense to describe techniques applied to structures at the human scale, most notably in the analysis of geographic data. The phrase is even sometimes used to refer to a specific technique in a single area of research, for example, to describe geostatistics.

The history of spatial analysis starts with early cartography, surveying and geography at the beginning of history, although the techniques of spatial analysis were not formalized until the later part of the twentieth century. Modern spatial analysis focuses on computer based techniques because of the large amount of data, the power of modern statistical and geographic information science (GIS) software, and the complexity of the computational modelling. Spatial analytic techniques have been developed in geography, biology, epidemiology, sociology, demography, statistics, geoinformatics, computer science, mathematics, and scientific modelling.

Complex issues arise in spatial analysis, many of which are neither clearly defined nor completely resolved, but form the basis for current research. The most fundamental of these is the problem of defining the spatial location of the entities being studied. For example, a study on human health could describe the spatial position of humans with a point placed where they live, or with a point located where they work, or by using a line to describe their weekly trips; each choice has dramatic effects on the techniques which can be used for the analysis and on the conclusions which can be obtained. Other issues in spatial analysis include the limitations of mathematical knowledge, the assumptions required by existing statistical techniques, and problems in computer based calculations.

Classification of the techniques of spatial analysis is difficult because of the large number of different fields of research involved, the different fundamental approaches which can be chosen, and the many forms the data can take.

The History of Spatial Analysis

Spatial analysis can perhaps be considered to have arisen with the early attempts at cartography and surveying but many fields have contributed to its rise in modern form. Biology contributed through botanical studies of global plant distributions and local plant locations, ethological studies of animal movement, landscape ecological studies of vegetation blocks, ecological studies of spatial population dynamics, and the study of biogeography. Epidemiology contributed with early work on disease mapping, notably John Snow's work mapping an outbreak of cholera, with research on mapping the spread of disease and with locational studies for health care delivery. Statistics has contributed greatly through work in spatial statistics. Economics has contributed notably through spatial econometrics. Geographic information system is currently a major contributor due to the importance of geographic software in the modern analytic toolbox. Remote sensing has contributed extensively in morphometric and clustering analysis. Computer science has contributed extensively through the study of algorithms, notably in computational geometry. Mathematics continues to provide the fundamental tools for analysis and to reveal the complexity of the spatial realm, for example, with recent work on fractals and scale invariance. Scientific modelling provides a useful framework for new approaches.

Fundamental Issues in Spatial Analysis

Spatial analysis confronts many fundamental issues in the definition of its objects of study, in the construction of the analytic operations to be used, in the use of computers for analysis, in the limitations and particularities of the analyses which are known, and in the presentation of analytic results. Many of these issues are active subjects of modern research.

Common errors often arise in spatial analysis, some due to the mathematics of space, some due to the particular ways data are presented spatially, some due to the tools which are available. Census data, because it protects individual privacy by aggregating data into local units, raises a number of statistical issues. Computer software can easily calculate the lengths of the lines which it defines but these may have no inherent meaning in the real world, as was shown for the coastline of Britain.

These problems represent one of the greatest dangers in spatial analysis because of the inherent power of maps as media of presentation. When results are presented as maps, the presentation combines the spatial data which is generally very accurate with analytic results which may be grossly inaccurate. Some of these issues are discussed at length in the book *How to Lie with Maps*

Spatial Characterization

The definition of the spatial presence of an entity constrains the possible analysis which can be applied to that entity and influences the final conclusions that can be reached. While this property is fundamentally true of all analysis, it is particularly important in spatial analysis because the tools to define and study entities favor specific characterizations of the entities being studied. Statistical techniques favor the spatial definition of objects as points because there are very few statistical techniques which operate directly on line, area, or volume elements. Computer tools favor the spatial definition of objects as homogeneous and separate elements because of the primitive nature of the computational structures available and the ease with which these primitive structures can be created.

There may also be arbitrary effects introduced by the spatial bounds or limits placed on the phenomenon or study area. This occurs since spatial phenomena may be unbounded or have ambiguous transition zones. This creates edge effects from ignoring spatial dependency or interaction outside the study area. It also imposes artificial shapes on the study area that can affect apparent spatial patterns such as the degree of clustering. A possible solution is similar to the sensitivity analysis strategy for the modifiable areal unit problem, or MAUP: change the limits of the study area and compare the results of the analysis under each realization. Another possible solution is to overbound the study area. It is also feasible to eliminate edge effects in spatial modelling and simulation by mapping the region to a boundless object such as a torus or sphere.

Spatial Dependency or Auto-correlation

A fundamental concept in geography is that nearby entities often share more similarities than entities which are far apart. This idea is often labeled 'Tobler's first law of geography' and may be summarized as "everything is related to everything else, but near things are more related than distant things".

Spatial dependency is the co-variation of properties within geographic space: characteristics at proximal locations appear to be correlated, either positively or negatively. There are at least three possible explanations. One possibility is there is a simple spatial correlation relationship: whatever is causing an observation in one location also causes similar observations in nearby locations.

For example, physical crime rates in nearby areas within a city tend to be similar due to factors such as socio-economic status, amount of policing and the built environment creating the opportunities for that kind of crime: the features that attract one criminal will also attract others. Another possibility is spatial causality: something at a given location directly influences the characteristics of nearby locations.

For example, the broken window theory of personal crime suggests that poverty, lack of maintenance and petty physical crime tends to breed more crime of this kind due to the apparent breakdown in order. A third possibility is spatial interaction: the movement of people, goods or information creates apparent relationships between locations. The "journey to crime" theory suggests that criminal activity occurs as a result of accessibility to a criminal's home, hangout or other key locations in his or her daily activities.

Spatial dependency leads to the spatial autocorrelation problem in statistics since, like temporal autocorrelation, this violates standard statistical techniques that assume independence among observations. For example, regression analyses that do not compensate for spatial dependency can have unstable parameter estimates and yield unreliable significance tests. Spatial regression models capture these relationships and do not suffer from these weaknesses. It is also appropriate to view spatial dependency as a source of information rather than something to be corrected.

Locational effects also manifest as spatial heterogeneity, or the apparent variation in a process with respect to location in geographic space. Unless a space is uniform and boundless, every location will have some degree of uniqueness relative to the other locations. This affects the spatial dependency relations and therefore the spatial process. Spatial heterogeneity means that overall parameters estimated for the entire system may not adequately describe the process at any given location.

Scaling

One of these issues is a simple issue of language. Different fields use "large scale" and "small scale" to mean the opposite things, for example, cartographers referring to the mathematical size of the scale ratio, 1/24000 being 'larger' than 1/100000, while landscape ecologists long referred to the extent of their study areas, with continents being 'larger' than forests.

The more fundamental issue of scale requires ensuring that the conclusion of the analysis does not depend on any arbitrary scale. Landscape ecologists failed to do this for many years and for a long time characterized landscape elements with quantitative metrics which depended on the scale at which they were measured. They eventually developed a series of scale invariant metrics.

Sampling

Spatial sampling involves determining a limited number of locations in geographic space for faithfully measuring phenomena that are subject to dependency and heterogeneity. Dependency suggests that since one location can predict the value of another location, we do not need observations in both places. But heterogeneity suggests that this relation can change across space, and therefore we cannot trust an observed degree of dependency beyond a region that may be small. Basic spatial sampling schemes include random, clustered and systematic. These basic schemes can be applied at multiple levels in a designated spatial hierarchy (e.g., urban area, city, neighborhood). It is also possible to exploit ancillary data, for example, using property values as a guide in a spatial sampling scheme to measure educational attainment and income. Spatial models such as autocorrelation statistics, regression and interpolation can also dictate sample design.

Common Errors in Spatial Analysis

The fundamental issues in spatial analysis lead to numerous problems in analysis including bias, distortion and outright errors in the conclusions reached. These issues are often interlinked but various attempts have been made to separate out particular issues from each other.

Length

In a paper by Benoit Mandelbrot on the coastline of Britain it was shown that it is inherently nonsensical to discuss certain spatial concepts despite an inherent presumption of the validity of the concept. Lengths in ecology depend directly on the scale at which they are measured and experienced. So while surveyors commonly measure the length of a river, this length only has meaning in the context of the relevance of the measuring technique to the question under study.

Locational Fallacy

The locational fallacy refers to error due to the particular spatial characterization chosen for the elements of study, in particular choice of placement for the spatial presence of the element.

Spatial characterizations may be simplistic or even wrong. Studies of humans often reduce the spatial existence of humans to a single point, for instance their home address. This can easily lead to poor analysis, for example, when considering disease transmission which can happen at work or at school and therefore far from the home.

The spatial characterization may implicitly limit the subject of study. For example, the spatial analysis of crime data has recently become popular but these studies can only describe the particular kinds of crime which can be described spatially. This leads to many maps of assault but not to any maps of embezzlement with political consequences in the conceptualization of crime and the design of policies to address the issue.

Atomic Fallacy

This describes errors due to treating elements as separate 'atoms' outside of their spatial context.

Ecological Fallacy

The ecological fallacy describes errors due to performing analyses on aggregate data when trying to reach conclusions on the individual units. It is closely related to the modifiable areal unit problem.

Modifiable Areal Unit Problem

The modifiable areal unit problem (MAUP) is an issue in the analysis of spatial data arranged in zones, where the conclusion depends on the particular shape or size of the zones used in the analysis.

Spatial analysis and modelling often involves aggregate spatial units such as census tracts or traffic analysis zones. These units may reflect data collection and/or modelling convenience rather than homogeneous, cohesive regions in the real world. The spatial units are therefore arbitrary or modifiable and contain artifacts related to the degree of spatial aggregation or the placement of boundaries.

The problem arises because it is known that results derived from an analysis of these zones depends directly on the zones being studied. It has been shown that the aggregation of point data into zones of different shapes and sizes can lead to opposite conclusions. More detail is available at the modifiable areal unit problem topic entry.

Solutions to the Fundamental Issues

Geographic Space

A mathematical space exists whenever we have a set of observations and quantitative measures of their attributes. For example, we can represent individuals' income or years of education within a coordinate system where the location of each individual can be specified with respect to both dimensions. The distances between individuals within this space is a quantitative measure of their differences with respect to income and education. However, in spatial analysis we are concerned with specific types of mathematical spaces, namely, geographic space. In geographic space, the observations correspond to locations in a spatial measurement framework that captures their proximity in the real world. The locations in a spatial measurement framework often represent locations on the surface of the Earth, but this is not strictly necessary. A spatial measurement framework can also capture proximity with respect to, say, interstellar space or within a biological entity such as a liver. The fundamental tenet is Tobler's First Law of Geography: if the interrelation between entities increases with proximity in the real world, then representation in geographic space and assessment using spatial analysis techniques are appropriate.

The Euclidean distance between locations often represents their proximity, although this is only one possibility. There are an infinite number of distances in addition to Euclidean that can support quantitative analysis. For example, "Manhattan" (or "Taxicab") distances where movement is restricted to paths parallel to the axes can be more meaningful than Euclidean distances in urban settings. In addition to distances, other geographic relationships such as connectivity (e.g., the existence or degree of shared borders) and direction can also influence the relationships among entities. It is also possible to compute minimal cost paths across a cost surface; for example, this can represent proximity among locations when travel must occur across rugged terrain.

Types of Spatial Analysis

Spatial data comes in many varieties and it is not easy to arrive at a system of classification that is simultaneously exclusive, exhaustive, imaginative, and satisfying. — G. Upton & B. Fingelton

Spatial Autocorrelation

Spatial autocorrelation statistics measure and analyze the degree of dependency among observations in a geographic space. Classic spatial autocorrelation statistics include Moran's *I* and Geary's *C*. These require measuring a spatial weights matrix that reflects the intensity of the geographic relationship between observations in a neighborhood, e.g., the distances between neighbors, the lengths of shared border, or whether they fall into a specified directional class such as "west." Classic spatial autocorrelation statistics compare the spatial weights to the covariance relationship at pairs of locations. Spatial autocorrelation that is more positive than expected from random indicate the clustering of similar values across geographic space, while significant negative spatial autocorrelation indicates that neighboring values are more dissimilar than expected by chance, suggesting a spatial pattern similar to a chess board.

Spatial autocorrelation statistics such as Moran's *I* and Geary's *C* are global in the sense that they estimate the overall degree of spatial autocorrelation for a dataset. The possibility of spatial heterogeneity suggests that the estimated degree of autocorrelation may vary significantly across geographic space. Local spatial autocorrelation statistics provide estimates disaggregated to the level of the spatial analysis units, allowing assessment of the dependency relationships across space. *G* statistics compare neighborhoods to a global average and identify local regions of strong autocorrelation. Local versions of the *I* and *C* statistics are also available.

Spatial Interpolation

Spatial interpolation methods estimate the variables at unobserved locations in geographic space based on the values at observed locations. Basic methods include inverse distance weighting: this attenuates the variable with decreasing proximity from the observed location. Kriging is a more sophisticated method that interpolates across space according to a spatial lag relationship that has both systematic and random components.

This can accommodate a wide range of spatial relationships for the hidden values between observed locations. Kriging provides optimal estimates given the hypothesized lag relationship, and error estimates can be mapped to determine if spatial patterns exist.

Spatial Regression

Spatial regression methods capture spatial dependency in regression analysis, avoiding statistical problems such as unstable parameters and unreliable significance tests, as well as providing information on spatial relationships among the variables involved.

Depending on the specific technique, spatial dependency can enter the regression model as relationships between the independent variables and the dependent, between the dependent variables and a spatial lag of itself, or in the error terms.

Geographically weighted regression (GWR) is a local version of spatial regression that generates parameters disaggregated by the spatial units of analysis. This allows assessment of the spatial heterogeneity in the estimated relationships between the independent and dependent variables.

Spatial Interaction

Spatial interaction or "gravity models" estimate the flow of people, material or information between locations in geographic space. Factors can include origin propulsive variables such as the number of commuters in residential areas, destination attractiveness variables such as the amount of office space in employment areas, and proximity relationships between the locations measured in terms such as driving distance or travel time.

In addition, the topological, or connective, relationships between areas must be identified, particularly considering the often conflicting relationship between distance and topology; for example, two spatially close neighborhoods may not display any significant interaction if they are separated by a highway. After specifying the functional forms of these relationships, the analyst can estimate model parameters using observed flow data and standard estimation techniques such as ordinary least squares or maximum likelihood.

Competing destinations versions of spatial interaction models include the proximity among the destinations (or origins) in addition to the origin-destination proximity; this captures the effects of destination (origin) clustering on flows. Computational methods such as artificial neural networks can also estimate spatial interaction relationships among locations and can handle noisy and qualitative data.

Simulation and Modelling

Spatial interaction models are aggregate and top-down: they specify an overall governing relationship for flow between locations. This characteristic is also shared by urban models such as those based on mathematical programming, flows among economic sectors, or bid-rent theory. An alternative modelling perspective is to represent the system at the highest possible level of disaggregation and study the bottom-up emergence of complex patterns and relationships from behaviour and interactions at the individual level....

Complex adaptive systems theory as applied to spatial analysis suggests that simple interactions among proximal entities can lead to intricate, persistent and functional spatial entities at aggregate levels. Two fundamentally spatial simulation methods are cellular automata and agent-based modelling. Cellular automata modelling imposes a fixed spatial framework such as grid cells and specifies rules that dictate the state of a cell based on the states of its neighboring cells. As time progresses, spatial patterns emerge as cells change states based on their neighbors; this alters the conditions for future time periods.

For example, cells can represent locations in an urban area and their states can be different types of land use. Patterns that can emerge from the simple interactions of local land uses include office districts and urban sprawl. Agent-based modelling uses software entities (agents) that have purposeful behaviour (goals) and can react, interact and modify their environment while seeking their objectives.

Unlike the cells in cellular automata, agents can be mobile with respect to space. For example, one could model traffic flow and dynamics using agents representing individual vehicles that try to minimize travel time between specified origins and destinations. While pursuing minimal travel times, the agents must avoid collisions with other vehicles also seeking to minimize their travel times. Cellular automata and agent-based modelling are complementary modelling strategies. They can be integrated into a common geographic automata system where some agents are fixed while others are mobile.

Geographic Information Science and Spatial Analysis

Geographic information systems (GIS) and the underlying geographic information science that advances these technologies have a strong influence on spatial analysis. The increasing ability to capture and handle geographic data means that spatial analysis is occurring within increasingly data-rich environments. Geographic data capture systems include remotely sensed imagery, environmental monitoring systems such as intelligent transportation systems, and location-aware technologies such as mobile devices that can report location in near-real time. GIS provide platforms for managing these data, computing spatial relationships such as distance, connectivity and directional relationships between spatial units, and visualizing both the raw data and spatial analytic results within a cartographic context.

Geovisualization (GVis) combines scientific visualization with digital cartography to support the exploration and analysis of geographic data and information, including the results of spatial analysis or simulation. GVis leverages the human orientation towards visual information processing in the exploration, analysis and communication of geographic data and information. In contrast with traditional cartography, GVis is typically three or four-dimensional (the latter including time) and user-interactive.

Geographic knowledge discovery (GKD) is the human-centred process of applying efficient computational tools for exploring massive spatial databases. GKD includes geographic data mining, but also encompasses related activities such as data selection, data cleaning and pre-processing, and interpretation of results. GVis can also serve a central role in the GKD process.

GKD is based on the premise that massive databases contain interesting (valid, novel, useful and understandable) patterns that standard analytical techniques cannot find. GKD can serve as a hypothesis-generating process for spatial analysis, producing tentative patterns and relationships that should be confirmed using spatial analytical techniques.

Spatial Decision Support Systems (sDSS) take existing spatial data and use a variety of mathematical models to make projections into the future. This allows urban and regional planners to test intervention decisions prior to implementation.

Geodemographic Segmentation

In marketing, Geodemographic segmentation is a multivariate statistical classification technique for discovering whether the individuals of a population fall into different groups by making quantitative comparisons of multiple characteristics with the assumption that the differences within any group should be less than the differences between groups.

Technologies Employed

The information technologies employed in geodemographic segmentation include geographic information system and database management software.

- Geographic information system: a business tool for interpreting data that consists of a demographic database, digitized maps, a computer and software.
- Database management software: a computer program in which data are captured on the computer, updated, maintained and organized for effective use and manipulation of data.

Principles

Geodemographic segmentation is based on two simple principles:

- People who live in the same neighborhood are more likely to have similar characteristics than are two people chosen at random.
- Neighborhoods can be categorized in terms of the characteristics of the population which they contain. Any two neighborhoods can be placed in the same category, i.e., they contain similar types of people, even though they are widely separated.

Geodemographic Segmentation Systems

Famous geodemographic segmentation systems are Prizm (US), Tapestry (US), CAMEO (UK), ACORN (UK) and MOSAIC (UK) system. New systems targeting subgroups of the population are also emerging. For example, Segmentos examines the geodemographic lifestyles of Hispanics in the United States.

CAMEO System

The CAMEO Classifications is a set of consumer classifications that are used internationally by organisations as part of their sales, marketing and network planning strategies. CAMEO UK has been built at postcode level and classifies over 60 million British consumers. It has been built to accurately segment the British market into 57 distinct neighbourhood types and 10 key marketing segments. CAMEO was developed and is maintained by Eurodirect.

ACORN System

A Classification Of Residential Neighborhoods (ACORN) system is conducted by Consolidated Analysis Centers Incorporated (CACI).

It is the first and leading geodemographic tool to identify and understand the UK population and the demand for products and services. ACORN categorizes all 1.9 million UK postcodes using over 125 demographic statistics within England, Scotland, Wales and Northern Ireland and employing 287 lifestyle variables. The classification system of ACORN contains 56 types of household under the 14 groups in 5 categories.

MOSAIC System

Mosaic UK is Experian's people classification system. Originally created by Prof Richard Webber (visiting Professor of Geography at Kings College University, London) in association with Experian. The latest version of Mosaic was released in 2009. It classifies the UK population into 15 main socio-economic groups and, within this, 67 different types.

Mosaic UK is part of a family of Mosaic classifications that covers 29 countries including most of Western Europe, the United States, Australia and the Far East.

Mosaic Global is Experian's global consumer classification tool. It is based on the simple proposition that the world's cities share common patterns of residential segregation. Mosaic Global is a consistent segmentation system that covers over 400 million of the world's households using local data from 29 countries. It has identified 10 types of residential neighbourhood that can be found in each of the countries. These systems are consisted of the different types of businesses.

geoSmart System

In Australia, geoSmart is a geodemographic segmentation system based on the principle that people with similar demographic profiles and lifestyles tend to live near each other. It is developed by an Australian supplier of geodemographic solutions, RDA Research.

geoSmart geodemographic segments are produced from the Australian Census (Australian Bureau of Statistics) demographic measures and modeled characteristics, and the system is updated for recent household growth. The clustering creates a single segment code that is represented by a descriptive statement or a thumbnail sketch.

In Australia, geoSmart is mainly used for database segmentation, customer acquisition, trade area profiling and letterbox targeting, although it can be used in a broad range of other applications.

Suitability Analysis

Suitability analysis is a GIS-based process used to determine the appropriateness of a given area for a particular use. The basic premise of suitability analysis is that each aspect of the landscape has intrinsic characteristics that are in some degree either suitable or unsuitable for the activities being planned. Suitability is determined through systematic, multi-factor analysis of the different aspect of the terrain. Model inputs include a variety of physical, cultural, and economic factors. The results are often displayed on a map that is used to highlight areas from high to low suitability.

A suitability model typically answers the question, "Where is the best location?" — whether it involves finding the best location for a new road or pipeline, a new housing development, or a retail store. For instance, a commercial developer building a new retail store may take into consideration distance to major highways and any competitors' stores, then combine the results with land use, population density, and consumer spending data to decide on the best location for that store.

Applications

- Land use analysis: Most jurisdictions use land suitability analysis for site selection, impact studies, and land use planning.
- Retail site selection: Suitability analysis is critical for both marketing and merchandising purposes, as well as for choosing new retail locations.
- Agriculture
- Defence
- Crime analysis

Geospatial Predictive Modelling

Geospatial predictive modelling is conceptually rooted in the principle that the occurrences of events being modeled are limited in distribution. Occurrences of events are neither uniform nor random in distribution – there are spatial environment factors (infrastructure, sociocultural, topographic, etc.) that constrain and influence where the locations of events occur.

Geospatial predictive modelling attempts to describe those constraints and influences by spatially correlating occurrences of historical geospatial locations with environmental factors that represent those constraints and influences. Geospatial predictive modelling is a process for analyzing events through a geographic filter in order to make statements of likelihood for event occurrence or emergence.

Predictive Models

There are two broad types of geospatial predictive models: deductive and inductive.

Deductive Method

The deductive method relies on qualitative data or a subject matter expert (SME) to describe the relationship between event occurrences and factors that describe the environment. As a result, the deductive process generally will rely on more subjective information. The means that the modeler could potentially be limiting the model by only inputting a number of factors that the human brain can comprehend.

An example of a deductive model is as follows: Sets of events are typically found ...

- Between 100 and 700 meters from airports.
- In the grassland land cover category.
- At elevations between 1000 and 1500 meters.

In this deductive model, high suitability locations for the set of events are constrained and influenced by non-empirically calculated spatial ranges for airports, land cover, and elevation: lower suitability areas would be everywhere else. The accuracy and detail of the deductive model is limited by the depth of qualitative data inputs to the model.

Inductive Method

The inductive method relies on the empirically-calculated spatial relationship between historical or known event occurrence locations and factors that make up the environment (infrastructure, socio-culture, topographic, etc.). Each event occurrence is plotted in geographic space and a quantitative relationship is defined between the event occurrence and the factors that make up the environment. The advantage of this method is that software can be developed to empirically discover– harnessing the speed of computers, which is crucial when hundreds of factors are involved – both known and unknown correlations between factors and events. Those quantitative relationship values are then processed by a statistical function to find spatial patterns that define high and low su

4

Comparisons Techniques in Geography

As spatial interrelationships are fundamental to geography, maps are a key tool. Classical cartography has been joined by a more modern approach to geographical analysis, computer-based geographic information systems (GIS).

Geographers use four interrelated approaches:

- Systematic-Groups geographical knowledge into categories that can be explored globally.
- Regional-Examines systematic relationships between categories for a specific region or location on the planet.
- Descriptive-Simply specifies the locations of features and populations.
- Analytical-Asks *why* we find features and populations in a specific geographic area.

Techniques used by geographers include, but are not limited to:

Cartography

Cartography studies the representation of the Earth's surface with abstract symbols (map making). Although other subdisciplines of geography rely on maps for presenting their analyses, the actual making of maps is abstract enough to be regarded separately. Cartography has grown from a collection of drafting techniques into an actual science.

Cartographers must learn cognitive psychology and ergonomics to understand which symbols convey information about the Earth most effectively, and behavioral psychology to induce the readers of their maps to act on the information. They must learn geodesy and fairly advanced mathematics to understand how the shape of the Earth affects the distortion of map symbols projected onto a flat surface for viewing. It can be said, without much controversy, that cartography is the seed from which the larger field of geography grew. Most geographers will cite a childhood fascination with maps as an early sign they would end up in the field.

Geographic Information Systems

Geographic information systems (GIS) deal with the storage of information about the Earth for automatic retrieval by a computer, in an accurate manner appropriate to the information's purpose. In addition to all of the other subdisciplines of geography, GIS specialists must understand computer science and database systems. GIS has revolutionized the field of cartography; nearly all mapmaking is now done with the assistance of some form of GIS software. GIS also refers to the science of using GIS software and GIS techniques to represent, analyze and predict spatial relationships. In this context, GIS stands for Geographic Information Science.

Remote Sensing

Remote sensing can be defined as the art and science of obtaining information about Earth features from measurements made at a distance. Remotely sensed data comes in many forms such as satellite imagery, aerial photography and data obtained from hand-held sensors. Geographers increasingly use remotely sensed data to obtain information about the Earth's land surface, ocean and atmosphere because it: a) supplies objective information at a variety of spatial scales (local to global), b) provides a synoptic view of the area of interest, c) allows access to distant and/or inaccessible sites, d) provides spectral information outside the visible portion of the electromagnetic spectrum, and e) facilitates studies of how features/areas change over time. Remotely sensed data may be analyzed either independently of, or in conjunction with, other digital data layers (e.g., in a Geographic Information System).

Geostatistics

Geostatistics deal with quantitative data analysis, specifically the application of statistical methodology to the exploration of geographic phenomena. Geostatistics is used extensively in a variety of fields including: hydrology, geology, petroleum exploration, weather analysis, urban planning, logistics, and epidemiology. The mathematical basis for geostatistics derives from cluster analysis, discriminant analysis, and non-parametric statistical tests, and a variety of other subjects. Applications of geostatistics rely heavily on Geographic Information Systems, particularly for the interpolation (estimate) of unmeasured points.

Ethnography

Geographic qualitative methods, or ethnographical; research techniques, are used by human geographers. In cultural geography there is a tradition of employing qualitative research techniques also used in anthropology and sociology. Participant observation and in-depth interviews provide human geographers with qualitative data.

Comparing Correlations Between Different Areas

There are different situations in which one might want to compare correlations/regressions:

1. Comparing the correlations/regressions between variables x and y in different groups of subjects
2. Comparing correlations/regressions within a single group of subjects
   (a) Correlation/regressions between variable j and k vs. correlation between variables j and h.
   (b) Correlation/regression between variable j and k vs. correlation between variables h and m.

Note that when we have two variables (x and y) the significance test for the correlation between them gives identical results to the significance test on the regression coefficient.

However, comparing two correlations is not the same as comparing two regressions. The regression coefficient is the slope of the best-fitting line relating the dependent variable to the predictor. The correlation indexes the degree of spread around that line.

So it is entirely possible for two regression lines to have identical slopes (same regression) but the data to be tightly clustered around one regression line (high correlation) and significantly less tightly clustered around the second line (i.e., different correlations). Conversely, the two regression lines may have very different slopes but the size of the correlation (degree of closeness to the line) may be very similar. This is illustrated in the SPSS outputs on the following pages. These data are available in a dataset (on web and J drive) called regs_and_corrs.sav. There are two groups of 70 subjects (group 1 and 2), and 3 variables (*indepvar*, *depvar1*, & *depvar2*). You might try some of the analyses, explained in the notes below, using these data. In particular:

- compare the correlation between *depvar1* and *indepvar* for group 1 with that for group 2;
- compare the regression B coefficient for *indepvar* in predicting *depvar1* in group 1 with that for group 2;
- repeat the above two analyses for the relationship between *depvar2* and *indepvar*
- in group 1 only compare the correlation between *depvar1* and *indepvar* with that between *depvar2* and *indepvar*;
- repeat the above for group 2 only.

In doing the above analyses you will find it useful to employ the syntax commands including in the file compcorr_syntax.sps (available on the J drive and via the web). You will probably find it useful to create your own syntax for computing the Fisher Z statistic.

Correlation and Dependence

In statistics, correlation and dependence are any of a broad class of statistical relationships between two or more random variables or observed data values.

Familiar examples of dependent phenomena include the correlation between the physical statures of parents and their offspring, and the correlation between the demand for a product and its price. Correlations are useful because they can indicate a predictive relationship that can be exploited in practice. For example, an electrical utility may produce less power on a mild day based on the correlation between electricity demand and weather. Correlations can also suggest possible causal, or mechanistic relationships; however, statistical dependence is not sufficient to demonstrate the presence of such a relationship.

Formally, *dependence* refers to any situation in which random variables do not satisfy a mathematical condition of probabilistic independence. In general statistical usage, *correlation* or *co-relation* can refer to any departure of two or more random variables from independence, but most commonly refers to a more specialized type of relationship between mean values. There are several *correlation coefficients*, often denoted *ñ* or *r*, measuring the degree of correlation. The most common of these is the Pearson correlation coefficient, which is mainly sensitive to a linear relationship between two variables. Other correlation coefficients have been developed to be more robust than the Pearson correlation, or more sensitive to nonlinear relationships.

Pearson's Product-moment Coefficient

The most familiar measure of dependence between two quantities is the Pearson product-moment correlation coefficient, or "Pearson's correlation." It is obtained by dividing the covariance of the two variables by the product of their standard deviations. Karl Pearson developed the coefficient from a similar but slightly different idea by Francis Galton.

The population correlation coefficient ñ$_{X,Y}$ between two random variables X and Y with expected values ì$_X$ and ì$_Y$ and standard deviations ó$_X$ and ó$_Y$ is defined

$$\rho_{X,Y} = \mathrm{corr}(X, Y) = \frac{\mathrm{cov}(X, Y)}{\sigma_X \sigma_Y} = \frac{E[(X - \mu_X)(Y - \mu_Y)]}{\sigma_X \sigma_Y}:$$

Where E is the expected value operator, *cov* means covariance, and, *corr* a widely used alternative notation for

Pearson's correlation. The Pearson correlation is defined only if both of the standard deviations are finite and both of them are nonzero. It is a corollary of the Cauchy–Schwarz inequality that the correlation cannot exceed 1 in absolute value. The correlation coefficient is symmetric: corr(*X*,*Y*) = corr(*Y*,*X*).

The Pearson correlation is +1 in the case of a perfect positive (increasing) linear relationship, "1 in the case of a perfect decreasing (negative) linear relationship, and some value between "1 and 1 in all other cases, indicating the degree of linear dependence between the variables. As it approaches zero there is less of a relationship. The closer the coefficient is to either "1 or 1, the stronger the correlation between the variables.

If the variables are independent, Pearson's correlation coefficient is 0, but the converse is not true because the correlation coefficient detects only linear dependencies between two variables. For example, suppose the random variable *X* is symmetrically distributed about zero, and $Y = X^2$. Then *Y* is completely determined by *X*, so that *X* and *Y* are perfectly dependent, but their correlation is zero; they are uncorrelated. However, in the special case when *X* and *Y* are jointly normal, uncorrelatedness is equivalent to independence.

If we have a series of *n* measurements of *X* and *Y* written as x_i and y_i where i = 1, 2,..., *n*, then the *sample correlation coefficient*, can be used to estimate the population Pearson correlation *r* between *X* and *Y*. The sample correlation coefficient is written:

$$r_{xy} = \frac{\sum_{i=1}^{n}(x_i - \bar{x})(y_i - \bar{y})}{(n-1)s_x s_y}:$$

where x and y are the sample means of *X* and *Y*, s_x and s_y are the sample standard deviations of *X* and *Y*.

This can also be written as:

$$r_{xy} = \frac{\sum x_i y_i - n\bar{x}\bar{y}}{(n-1)s_x s_y} = \frac{n\sum x_i y_i - \sum x_i \sum y_i}{\sqrt{n\sum x_i^2 - (\sum x_i)^2}\ \sqrt{n\sum y_i^2 - (\sum y_i)^2}}.$$

Rank Correlation Coefficients

Rank correlation coefficients, such as Spearman's rank correlation coefficient and Kendall's rank correlation coefficient (ô) measure the extent to which, as one variable increases, the other variable tends to increase, without requiring that increase to be represented by a linear relationship. If, as the one variable increase, the other *decreases*, the rank correlation coefficients will be negative. It is common to regard these rank correlation coefficients as alternatives to Pearson's coefficient, used either to reduce the amount of calculation or to make the coefficient less sensitive to non-normality in distributions. However, this view has little mathematical basis, as rank correlation coefficients measure a different type of relationship than the Pearson product-moment correlation coefficient, and are best seen as measures of a different type of association, rather than as alternative measure of the population correlation coefficient.

To illustrate the nature of rank correlation, and its difference from linear correlation, consider the following four pairs of numbers (x, y):

(0, 1), (10, 100), (101, 500), (102, 2000).

As we go from each pair to the next pair x increases, and so does y. This relationship is perfect, in the sense that an increase in x is *always* accompanied by an increase in y. This means that we have a perfect rank correlation, and both Spearman's and Kendall's correlation coefficients are 1, whereas in this example Pearson product-moment correlation coefficient is 0.456, indicating that the points are far from lying on a straight line.

In the same way if y always *decreases* when x *increases*, the rank correlation coefficients will be "1, while the Pearson product-moment correlation coefficient may or may not be close to-1, depending on how close the points are to a straight line. Although in the extreme cases of perfect rank correlation the two coefficients are both equal (being both +1 or both "1) this is not in general so, and values of the two coefficients cannot meaningfully be compared. For example, for the three pairs (1, 1) (2, 3) (3, 2) Spearman's coefficient is 1/2, while Kendall's coefficient is 1/3.

Other Measures of Dependence among Random Variables

The information given by a correlation coefficient is not enough to define the dependence structure between random variables. The correlation coefficient completely defines the dependence structure only in very particular cases, for example when the distribution is a multivariate normal distribution. In the case of elliptic distributions it characterizes the (hyper-)ellipses of equal density, however, it does not completely characterize the dependence structure (for example, a multivariate t-distribution's degrees of freedom determine the level of tail dependence).

To get a measure for more general dependencies in the data (also nonlinear) it is better to use the correlation ratio which is able to detect almost any functional dependency, or the entropy-based mutual information/total correlation which is capable of detecting even more general dependencies. The latter are sometimes referred to as multi-moment correlation measures, in comparison to those that consider only 2nd moment (pairwise or quadratic) dependence.

The polychoric correlation is another correlation applied to ordinal data that aims to estimate the correlation between theorised latent variables.

One way to capture a more complete view of dependence structure is to consider a copula between them.

Sensitivity to the Data Distribution

The degree of dependence between variables X and Y should not depend on the scale on which the variables are expressed. Therefore, most correlation measures in common use are invariant to location and scale transformations of the marginal distributions. That is, if we are analyzing the relationship between X and Y, most correlation measures are unaffected by transforming X to $a + bX$ and Y to $c + dY$, where a, b, c, and d are constants. This is true of most correlation statistics as well as their population analogues. Some correlation statistics, such as the rank correlation coefficient, are also invariant to monotone transformations of the marginal distributions of X

and/or Y. Most correlation measures are sensitive to the manner in which X and Y are sampled. Dependencies tend to be stronger if viewed over a wider range of values. Thus, if we consider the correlation coefficient between the heights of fathers and their sons over all adult males, and compare it to the same correlation coefficient calculated when the fathers are selected to be between 165 cm and 170 cm in height, the correlation will be weaker in the latter case.

Various correlation measures in use may be undefined for certain joint distributions of X and Y. For example, the Pearson correlation coefficient is defined in terms of moments, and hence will be undefined if the moments are undefined. Measures of dependence based on quantiles are always defined. Sample-based statistics intended to estimate population measures of dependence may or may not have desirable statistical properties such as being unbiased, or asymptotically consistent, based on the structure of the population from which the data were sampled.

Correlation Matrices

The correlation matrix of n random variables $X_1,..., X_n$ is the $n \times n$ matrix whose i,j entry is corr(X_i, X_j). If the measures of correlation used are product-moment coefficients, the correlation matrix is the same as the covariance matrix of the standardized random variables X_i/ó (X_i) for $i = 1, ..., n$. This applies to both the matrix of population correlations (in which case "ó " is the population standard deviation), and to the matrix of sample correlations (in which case "ó " denotes the sample standard deviation). Consequently, each is necessarily a positive-semidefinite matrix.

The correlation matrix is symmetric because the correlation between X_i and X_j is the same as the correlation between X_j and X_i.

Common Misconceptions

Correlation and Causality

The conventional dictum that "correlation does not imply causation" means that correlation cannot be used to infer a causal relationship between the variables. This dictum should not be taken to mean that correlations cannot indicate the potential existence of causal relations. However, the causes underlying the correlation, if any, may be indirect and unknown, and high correlations also overlap with identity relations, where no causal process exists. Consequently, establishing a correlation between two variables is not a sufficient condition to establish a causal relationship (in either direction). For example, one may observe a correlation between an ordinary alarm clock ringing and daybreak, though there is no causal relationship between these phenomena.

A correlation between age and height in children is fairly causally transparent, but a correlation between mood and health in people is less so. Does improved mood lead to improved health; or does good health lead to good mood; or both? Or does some other factor underlie both? In other words, a correlation can be taken as evidence for a possible causal relationship, but cannot indicate what the causal relationship, if any, might be.

Correlation and Linearity

The Pearson correlation coefficient indicates the strength of a linear relationship between two variables, but its value generally does not completely characterize their relationship. In particular, if the conditional mean of Y given X, denoted $E(Y|X)$, is not linear in X, the correlation coefficient will not fully determine the form of $E(Y|X)$.

The image on the right shows scatterplots of Anscombe's quartet, a set of four different pairs of variables created by Francis Anscombe. The four y variables have the same mean (7.5), standard deviation (4.12), correlation (0.816) and regression line ($y = 3 + 0.5x$). However, as can be seen on the plots, the distribution of the variables is very different. The first one (top left) seems to be distributed normally, and corresponds to what one would expect when considering two variables correlated and following the assumption of normality. The second one (top right) is not distributed normally; while an obvious relationship between the two variables can be observed, it is not linear. In this case the Pearson correlation coefficient does not indicate that there is an exact functional relationship: only the extent to which that relationship can be approximated by a linear relationship. In the third case (bottom left), the linear relationship is perfect, except for one outlier which exerts enough influence to lower the correlation coefficient from 1 to 0.816. Finally, the fourth example (bottom right) shows another example when one outlier is enough to produce a high correlation coefficient, even though the relationship between the two variables is not linear.

These examples indicate that the correlation coefficient, as a summary statistic, cannot replace the individual examination of the data. Note that the examples are sometimes said to demonstrate that the Pearson correlation assumes that the data follow a normal distribution, but this is not correct.

If a pair (X, Y) of random variables follows a bivariate normal distribution, the conditional mean $E(X|Y)$ is a linear function of Y, and the conditional mean $E(Y|X)$ is a linear function of X. The correlation coefficient r between X and Y, along with the marginal means and variances of X and Y, determines this linear relationship:

$$E(Y|X) = EY + r\sigma_y \frac{X - EX}{\sigma_x},$$

where EX and EY are the expected values of X and Y, respectively, and $ó_x$ and $ó_y$ are the standard deviations of X and Y, respectively.

Partial Correlation

If a population or data-set is characterised by more than two variables, a partial correlation coefficient measures the strength of dependence between a pair of variables that is not accounted for by the way in which they both change in response to variations in a selected subset of the other variables.

Networks and Classification

Remote Sensing

Remote sensing is the small-or large-scale acquisition of information of an object or phenomenon, by the use of either recording or real-time sensing device(s) that are wireless, or not in physical or intimate contact with the object (such as by way of aircraft, spacecraft, satellite, buoy, or ship). In practice, remote sensing is the stand-off collection through the use of a variety of devices for gathering information on a given object or area. Thus, Earth observation or weather satellite collection platforms, ocean and atmospheric observing weather buoy platforms, the monitoring of a parolee via an ultrasound identification system, Magnetic Resonance Imaging (MRI), Positron Emission Tomography (PET), X-radiation (X-RAY) and space probes are all examples of remote sensing. In modern usage, the term generally refers to the use of imaging sensor technologies including: instruments found in aircraft and spacecraft as well as those used in electrophysiology, and is distinct from other imaging-related fields such as medical imaging.

Overview

There are two main types of remote sensing: passive remote sensing and active remote sensing. Passive sensors detect natural radiation that is emitted or reflected by the object or surrounding area being observed. Reflected sunlight is the most common source of radiation measured by passive sensors. Examples of passive remote sensors include film photography, Infrared, charge-coupled devices, and radiometers. Active collection, on the other hand, emits energy in order to scan objects and areas whereupon a sensor then detects and measures the radiation that is reflected or backscattered from the target. RADAR is an example of active remote sensing where the time delay between emission and return is measured, establishing the location, height, speed and direction of an object.

Remote sensing makes it possible to collect data on dangerous or inaccessible areas. Remote sensing applications include monitoring deforestation in areas such as the Amazon Basin, the effects of climate change on glaciers and Arctic and Antarctic regions, and depth sounding of coastal and ocean depths. Military collection during the cold war made use of stand-off collection of data about dangerous border areas. Remote sensing also replaces costly and slow data collection on the ground, ensuring in the process that areas or objects are not disturbed.

Orbital platforms collect and transmit data from different parts of the electromagnetic spectrum, which in conjunction with larger scale aerial or ground-based sensing and analysis, provides researchers with enough information to monitor trends such as El Niño and other natural long and short term phenomena. Other uses include different areas of the earth sciences such as natural resource management, agricultural fields such as land usage and conservation, and national security and overhead, ground-based and stand-off collection on border areas.

Data Acquisition Techniques

The basis for multispectral collection and analysis is that of examined areas or objects that reflect or emit radiation that stand out from surrounding areas.

Applications of Remote Sensing Data

- Conventional radar is mostly associated with aerial traffic control, early warning, and certain large scale meteorological data. Doppler radar is used by local law enforcements' monitoring of speed limits and in enhanced meteorological collection such as wind speed and direction within weather systems. Other types of active collection includes plasmas in the ionosphere). Interferometric synthetic aperture radar is used to produce precise digital elevation models of large scale terrain.
- Laser and radar altimeters on satellites have provided a wide range of data. By measuring the bulges of water caused by gravity, they map features on the seafloor to a resolution of a mile or so. By measuring the height and wave-length of ocean waves, the altimeters measure wind speeds and direction, and surface ocean currents and directions.
- Light detection and ranging (LIDAR) is well known in examples of weapon ranging, laser illuminated homing of projectiles. LIDAR is used to detect and measure the

concentration of various chemicals in the atmosphere, while airborne LIDAR can be used to measure heights of objects and features on the ground more accurately than with radar technology. Vegetation remote sensing is a principle application of LIDAR.

- Radiometers and photometers are the most common instrument in use, collecting reflected and emitted radiation in a wide range of frequencies. The most common are visible and infrared sensors, followed by microwave, gamma ray and rarely, ultraviolet. They may also be used to detect the emission spectra of various chemicals, providing data on chemical concentrations in the atmosphere.
- Stereographic pairs of aerial photographs have often been used to make topographic maps by imagery and terrain analysts in trafficability and highway departments for potential routes.
- Simultaneous multi-spectral platforms such as Landsat have been in use since the 70's. These thematic mappers take images in multiple wavelengths of electro-magnetic radiation (multi-spectral) and are usually found on earth observation satellites, including (for example) the Landsat program or the IKONOS satellite. Maps of land cover and land use from thematic mapping can be used to prospect for minerals, detect or monitor land usage, deforestation, and examine the health of indigenous plants and crops, including entire farming regions or forests.
- Within the scope of the combat against desertification, remote sensing allows to follow-up and monitor risk areas in the long term, to determine desertification factors, to support decision-makers in defining relevant measures of environmental management, and to assess their impacts.

Geodetic

- Overhead geodetic collection was first used in aerial submarine detection and gravitational data used in military maps. This data revealed minute perturbations in the Earth's gravitational field (geodesy) that may be used to determine changes in the mass distribution of the Earth, which in turn may be used for geological or hydrological studies.

Acoustic and Near-acoustic

- Sonar: *passive sonar*, listening for the sound made by another object (a vessel, a whale etc); *active sonar*, emitting pulses of sounds and listening for echoes, used for detecting, ranging and measurements of underwater objects and terrain.
- Seismograms taken at different locations can locate and measure earthquakes (after they occur) by comparing the relative intensity and precise timing.

To coordinate a series of large-scale observations, most sensing systems depend on the following: platform location, what time it is, and the rotation and orientation of the sensor. High-end instruments now often use positional information from satellite navigation systems. The rotation and orientation is often provided within a degree or two with electronic compasses. Compasses can measure not just azimuth (i.e. degrees to magnetic north), but also altitude (degrees above the horizon), since the magnetic field curves into the Earth at different angles at different latitudes. More exact orientations require gyroscopic-aided orientation, periodically realigned by different methods including navigation from stars or known benchmarks.

Resolution impacts collection and is best explained with the following relationship: less resolution=less detail & larger coverage, More resolution=more detail, less coverage. The skilled management of collection results in cost-effective collection and avoid situations such as the use of multiple high resolution data which tends to clog transmission and storage infrastructure.

Data processing

Generally speaking, remote sensing works on the principle of the *inverse problem*. While the object or phenomenon of interest (the state) may not be directly measured, there exists some other variable that can be detected and measured (the observation), which may be related to the object of interest through the use of a data-derived computer model. The common analogy given to describe this is trying to determine the type of animal from its footprints. For example, while it is impossible to directly measure temperatures in the upper atmosphere, it is possible to measure the spectral emissions from a known chemical species (such as carbon dioxide) in that region. The frequency of the emission may then be related to the temperature in that region via various thermodynamic relations.

The quality of remote sensing data consists of its spatial, spectral, radiometric and temporal resolutions.

Spatial Resolution

The size of a pixel that is recorded in a raster image-typically pixels may correspond to square areas ranging in side length from 1 to 1,000 metres (3.3 to 3,300 ft).

Spectral Resolution

The wavelength width of the different frequency bands recorded-usually, this is related to the number of frequency bands recorded by the platform. Current Landsat collection is that of seven bands, including several in the infra-red spectrum, ranging from a spectral resolution of 0.07 to 2.1 ìm. The Hyperion sensor on Earth Observing-1 resolves 220 bands from 0.4 to 2.5 ìm, with a spectral resolution of 0.10 to 0.11 ìm per band.

Radiometric Resolution

The number of different intensities of radiation the sensor is able to distinguish. Typically, this ranges from 8 to 14 bits, corresponding to 256 levels of the gray scale and up to 16,384 intensities or "shades" of colour, in each band. It also depends on the instrument noise.

Temporal Resolution

The frequency of flyovers by the satellite or plane, and is only relevant in time-series studies or those requiring an averaged or mosaic image as in deforesting monitoring. This was first used by the intelligence community where repeated coverage revealed changes in infrastructure, the deployment of units or the modification/introduction of equipment. Cloud cover over a given area or object makes it necessary to repeat the collection of said location.

In order to create sensor-based maps, most remote sensing systems expect to extrapolate sensor data in relation to a reference point including distances between known points on the ground. This depends on the type of sensor used. For example, in conventional photographs, distances are accurate in the center of the image, with the distortion of measurements increasing the farther you get from the center. Another factor is that of the platen against which the film is pressed can cause severe errors when photographs are used to measure ground distances. The step in which this problem is resolved is called georeferencing, and involves computer-aided matching up of points in the image (typically 30 or more points per image) which is extrapolated with the use of an established benchmark, "warping" the image to produce accurate spatial data. As of the early 1990s, most satellite images are sold fully georeferenced.

In addition, images may need to be radiometrically and atmospherically corrected.

Radiometric correction gives a scale to the pixel values, e.g. the monochromatic scale of 0 to 255 will be converted to actual radiance values.

Atmospheric correction eliminates atmospheric haze by rescaling each frequency band so that its minimum value (usually realised in water bodies) corresponds to a pixel value of 0. The digitizing of data also make possible to manipulate the data by changing gray-scale values.

Interpretation is the critical process of making sense of the data. The first application was that of aerial photographic collection which used the following process; spatial measurement through the use of a light table in both conventional single or stereographic coverage, added skills such as the use of photogrammetry, the use of photomosaics, repeat coverage, Making use of objects' known dimensions in order to detect modifications. Image Analysis is the recently developed automated computer-aided application which is in increasing use. Object-Based Image Analysis (OBIA) is a sub-discipline of GIScience devoted to partitioning remote sensing (RS) imagery into meaningful image-objects, and assessing their characteristics through spatial, spectral and temporal scale.

Old data from remote sensing is often valuable because it may provide the only long-term data for a large extent of geography. At the same time, the data is often complex to interpret, and bulky to store. Modern systems tend to store the data digitally, often with lossless compression. The difficulty with this approach is that the data is fragile, the format may be archaic, and the data may be easy to falsify. One of the best systems for archiving data series is as computer-generated machine-readable ultrafiche, usually in typefonts such as OCR-B, or as digitized half-tone images. Ultrafiches survive well in standard libraries, with lifetimes of several centuries. They can be created, copied, filed and retrieved by automated systems. They are about as compact as archival magnetic media, and yet can be read by human beings with minimal, standardized equipment.

Data Processing Levels

To facilitate the discussion of data processing in practice, several processing "levels" were first defined in 1986 by NASA as part of its Earth Observing System and steadily adopted since then, both internally at NASA (e.g.,) and elsewhere (e.g.,); these definitions are:

Level	Description
0	Reconstructed, unprocessed instrument and payload data at full resolution, with any and all communications artifacts (e.g., synchronization frames, communications headers, duplicate data) removed.
1a	Reconstructed, unprocessed instrument data at full resolution, time-referenced, and annotated with ancillary information, including radiometric and geometric calibration coefficients and georeferencing parameters (e.g., platform ephemeris) computed and appended but not applied to the Level 0 data (or

if applied, in a manner that level 0 is fully recoverable from level 1a data).

1b Level 1a data that have been processed to sensor units (e.g., radar backscatter cross section, brightness temperature, etc.); not all instruments have Level 1b data; level 0 data is not recoverable from level 1b data.

2 Derived geophysical variables (e.g., ocean wave height, soil moisture, ice concentration) at the same resolution and location as Level 1 source data.

3 Variables mapped on uniform space-time grid scales, usually with some completeness and consistency (e.g., missing points interpolated, complete regions mosaicked together from multiple orbits, etc).

4 Model output or results from analyses of lower level data (i.e., variables that were not measured by the instruments but instead are derived from these measurements).

A Level 1 data record is the most fundamental (i.e., highest reversible level) data record that has significant scientific utility, and is the foundation upon which all subsequent data sets are produced. Level 2 is the first level that is directly usable for most scientific applications; its value is much greater than the lower levels. Level 2 data sets tend to be less voluminous than Level 1 data because they have been reduced temporally, spatially, or spectrally. Level 3 data sets are generally smaller than lower level data sets and thus can be dealt with without incurring a great deal of data handling overhead. These data tend to be generally more useful for many applications. The regular spatial and temporal organization of Level 3 datasets makes it feasible to readily combine data from different sources.

History

Beyond the primitive methods of remote sensing our earliest ancestors used (ex.: standing on a high cliff or tree to view the landscape), the modern discipline arose with the development of flight. The balloonist G. Tournachon (alias Nadar) made photographs of Paris from his balloon in 1858. Messenger pigeons, kites, rockets and unmanned balloons were also used for early images. With the exception of balloons, these first, individual images were not particularly useful for map making or for scientific purposes.

Systematic aerial photography was developed for military surveillance and reconnaissance purposes beginning in World War I and reaching a climax during the Cold War with the use of modified combat aircraft such as the P-51, P-38, RB-66 and the F-4C, or specifically designed collection platforms such as the U2/TR-1, SR-71, A-5 and the OV-1 series both in overhead and stand-off collection. A more recent development is that of increasingly smaller sensor pods such as those used by law enforcement and the military, in both manned and unmanned platforms. The advantage of this approach is that this requires minimal modification to a given airframe. Later imaging technologies would include Infra-red, conventional, doppler and synthetic aperture radar.

The development of artificial satellites in the latter half of the 20th century allowed remote sensing to progress to a global scale as of the end of the Cold War. Instrumentation aboard various Earth observing and weather satellites such as Landsat, the Nimbus and more recent missions such as RADARSAT and UARS provided global measurements of various data for civil, research, and military purposes. Space probes to other planets have also provided the opportunity to conduct remote sensing studies in extraterrestrial environments, synthetic aperture radar aboard the Magellan spacecraft provided detailed topographic maps of Venus, while instruments aboard SOHO allowed studies to be performed on the Sun and the solar wind, just to name a few examples.

Recent developments include, beginning in the 1960s and 1970s with the development of image processing of satellite imagery. Several research groups in Silicon Valley including NASA Ames Research Center, GTE and ESL Inc. developed Fourier transform techniques leading to the first notable enhancement of imagery data.

The introduction of online web services for easy access to remote sensing data in the 21st century (mainly low/medium-resolution images), like Google Earth, has made remote sensing more familiar to the big public and has popularized the science.

Remote Sensing Software

Remote Sensing data is processed and analyzed with computer software, known as a remote sensing application. A large number of proprietary and open source applications exist to process remote sensing data. According to an NOAA Sponsored Research by Global Marketing Insights, Inc. the most used applications among Asian academic groups involved in remote sensing are as follows: ERDAS 36% (ERDAS IMAGINE 25% & ERMapper 11%); ESRI 30%; ITT Visual Information Solutions ENVI 17%; MapInfo 17%. Among Western Academic respondents as follows: ESRI 39%, ERDAS IMAGINE 27%, MapInfo 9%, AutoDesk 7%, ITT Visual Information Solutions ENVI 17%. Other important Remote Sensing Software packages include: TNTmips from MicroImages, PCI Geomatica made by PCI Geomatics, the leading remote sensing software package in Canada, IDRISI from Clark Labs, Image Analyst from Intergraph, and the original object based image analysis software eCognition from Definiens. Dragon/ips is one of the oldest remote sensing packages still available, and is in some cases free. Open source remote sensing software includes GRASS GIS, QGIS, OSSIM, Opticks (software) and Orfeo toolbox.

Structure of Communication Networks

Communication networks based on serial data transmission are the platform of up-to-date automation systems. Whether this is office automation or automation of manufacturing or process plants, the task remains always the same, exchanging data between different devices or participants within a system. Communication networks provide a number of advantages over systems in which a point-to-point line enables only two participants to communicate with each other.

Classification of Communication Networks

Depending on the application, i.e. manufacturing, process, office or building automation, the communication tasks to be performed vary in complexity and are sometimes even contradictory. The use of only one communication network would therefore not yield optimum results. So the market offers very different networks and bus systems that are more or less tailored to a specific application.

A quite general classification criterion is the distance over which communication takes place. There are local networks, LANs (Local Area Networks), as well as widely distributed networks, WANs (Wide Area Networks). With LAN, emphasis is put on fast and powerful data exchange within a locally restricted area, whereas WAN must be able to transmit data on very different data media and over several thousand kilometers.

A telecommunications network is a collection of terminals, links and nodes which connect together to enable telecommunication between users of the terminals. Networks may use circuit switching or message switching. Each terminal in the network must have a unique address so messages or connections can be routed to the correct recipients. The collection of addresses in the network is called the address space.

The links connect the nodes together and are themselves built upon an underlying transmission network which physically pushes the message across the link.

Examples of telecommunications networks are:

- Computer network
- the Internet Network
- the telephone network
- the global Telex network
- the aeronautical ACARS network.

Messages and Protocols

Messages are generated by a sending terminal, then pass through the network of links and nodes until they arrive at the destination terminal. It is the job of the intermediate nodes to handle the messages and route them down the correct link toward their final destination. The messages consist of control (or signaling) and bearer parts which can be sent together or separately. The bearer part is the actual content that the user wishes to transmit (e.g. some encoded speech, or an email) whereas the control part instructs the nodes where and possibly how the message should be routed through the network. A large number of protocols have been developed over the years to specify how each different type of telecommunication network should handle the control and bearer messages to achieve this efficiently.

Telecommunication Network Components

All telecommunication networks are made up of five basic components that are present in each network environment regardless of type or use. These basic components include terminals, telecommunications processors, telecommunications channels, computers, and telecommunications control software.

- Terminals are the starting and stopping points in any telecommunication network environment. Any input or output device that is used to transmit or receive data can be classified as a terminal component.
- Telecommunications processors support data transmission and reception between terminals and computers by providing a variety of control and support functions. (i.e. convert data from digital to analog and back)
- Telecommunications channels are the way by which data is transmitted and received. Telecommunication channels are created through a variety of media of which the most popular include copper wires and coaxial cables. Fiber-optic cables are increasingly used to bring faster and more robust connections to businesses and homes.
- In a telecommunication environment computers are connected through media to perform their communication assignments.
- Telecommunications control software is present on all networked computers and is responsible for controlling network activities and functionality.

Early networks were built without computers, but late in the 20th century their switching centers were computerized or the networks replaced with computer networks.

Network Structure

In general, every telecommunications network conceptually consists of three parts, or planes (so called because they can be thought of as being, and often are, separate overlay networks):

- The Control Plane carries control information (also known as signalling).
- The Data Plane or User Plane carries the network's users' traffic.
- The Management Plane carries the operations and administration traffic required for network management.

Example of a Telecommunication Network: The TCP/IP Data Network

The data network is used extensively throughout the world to connect individuals and organizations. Data networks can be connected together to allow users seamless access to resources that are hosted outside of the particular provider they are connected to. The Internet is the best example of many data networks from different organizations all operating under a single address space.

Terminals attached to TCP/IP networks are addressed using IP addresses. There are different types of IP address, but the most common is IP Version 4. Each unique address consists of 4 integers between 0 and 255, usually separated by dots when written down, e.g. 82.131.34.56.

TCP/IP are the fundamental protocols that provide the control and routing of messages across the data network. There are many different network structures that TCP/IP can be used across to efficiently route messages, for example:

- wide area networks (WAN)
- metropolitan area networks (MAN)
- local area networks (LAN)

- Campus area networks (CAN)
- virtual private networks (VPN)

There are three features that differentiate MANs from LANs or WANs:

1. The area of the network size is between LANs and WANs. The MAN will have a physical area between 5 and 50 km in diameter.
2. MANs do not generally belong to a single organization. The equipment that interconnects the network, the links, and the MAN itself are often owned by an association or a network provider that provides or leases the service to others.
3. A MAN is a means for sharing resources at high speeds within the network. It often provides connections to WAN networks for access to resources outside the scope of the MAN.

Node (Networking)

In communication networks, a node (Latin *nodus*, 'knot') is a connection point, either a redistribution point or a communication endpoint (some terminal equipment). The definition of a node depends on the network and protocol layer referred to. A physical network node is an active electronic device that is attached to a network, and is capable of sending, receiving, or forwarding information over a communications channel. A passive distribution point such as a distribution frame is consequently not a node.

In network theory or graph theory, the term *node* refers to a point in a network topology at which lines intersect or branch

Computer Network Nodes

In data communication, a physical network node may either be a data circuit-terminating equipment (DCE) such as a modem, hub, bridge or switch; or a data terminal equipment (DTE) such as a digital telephone handset, a printer or a host computer, for example a router, a workstation or a server.

If the network in question is a LAN or WAN, every LAN or WAN node (that are at least data link layer devices) must have a MAC address.

Examples are computers, packet switches and ADSL modem (with Ethernet interface). Note that a hub constitutes a physical network node, but not a LAN node in this sense, since a hubbed network logically is a bus network. Analogously, a repeater or PSTN modem (with serial interface) are physical network nodes but not LAN nodes in this sense.

If the network in question is the Internet, many physical network nodes are host computers, also known as Internet nodes, identified by an IP address, and all hosts are physical network nodes. However, datalink layer devices such as switches, bridges and WLAN access points do not have an IP host address (except sometimes for administrative purposes), and are not considered as Internet nodes, but as physical network nodes or LAN nodes.

Telecommunication Network Nodes

In the fixed telephone network, a node may be a public or private telephone exchange, a remote concentrator or a computer providing some intelligent network service. In cellular communication, switching points and databases such as the Base station controller, Home Location Register, Gateway GPRS Support Node (GGSN) and Serving GPRS Support Node (SGSN) are examples of nodes. Cellular network base stations are not considered as nodes in this context.

In cable television systems (CATV), this term has assumed a broader context and is generally associated with a fiber optic node. This can be defined as those homes or businesses within a specific geographic area that are served from a common fiber optic receiver. A fiber optic node is generally described in terms of the number of "homes passed" that are served by that specific fiber node.

Distributed System Nodes

If the network in question is a distributed system, the nodes are clients, servers or peers. A peer may sometimes serve as client, sometimes server. In a peer-to-peer or overlay network, nodes that actively route data for the other networked devices as well as themselves are called supernodes.

End Node in Cloud Computing

Within a vast computer network, the individual computers on the periphery of the network, those that do not also connect other networks, and/or those that often attach transiently to one or more clouds are called end nodes. Typically, within the cloud computing construct, the individual user/customer computer that connects into one well-managed cloud is called an end node. Since these computers are a part of yet unmanaged by the cloud's host, they present significant risks to the entire cloud. This is called the End Node Problem. There are several means to remedy this problem but all require instilling trust in the end node computer.

Establishing the Boundaries

Borders define geographic boundaries of political entities or legal jurisdictions, such as governments, sovereign states, federated states and other subnational entities. Some borders—such as a state's internal administrative borders, or inter-state borders within the Schengen Area—are open and completely unguarded. Other borders are partially or fully controlled, and may be crossed legally only at designated border checkpoints. Some, mostly contentious, borders may even foster the setting up of buffer zones.

In the past many borders were not clearly defined lines, but were neutral zones called marchlands. This has been reflected in recent times with the neutral zones that were set up along part of Saudi Arabia's borders with Kuwait and Iraq (however, these zones no longer exist). In modern times the concept of a marchland has been replaced by that of the clearly defined and demarcated border. For the purposes of border control, airports and seaports are also classed as borders. Most countries have some form of border control to restrict or limit the movement of people, animals, plants, and goods into or out of the country. Under international law, each country is generally permitted to define the conditions that have to be met by a person to legally cross its borders by its own laws, and to prevent persons from crossing its border when this happens in violation of those laws.

In order to cross borders, the presentation of passports and visas or other appropriate forms of identity document is required by some legal orders. To stay or work within a country's borders aliens (foreign persons) may need special immigration documents or permits that authorise them to do so. Having such documents (i.e. visa and passport) however does not automatically guarantee that the alien will be allowed to cross to the other side of the border.

Moving goods across a border often requires the payment of excise tax, often collected by customs officials. Animals (and occasionally humans) moving across borders may need to go into quarantine to prevent the spread of exotic or infectious diseases.

Most countries prohibit carrying illegal drugs or endangered animals across their borders. Moving goods, animals or people illegally across a border, without declaring them, seeking permission, or deliberately evading official inspection constitutes smuggling.

Natural Borders

Natural borders are geographical features that present natural obstacles to communication and transport. Existing political borders are often a formalization of these historical, natural obstacles.

Some geographical features that often constitute natural borders are:

- Oceans: oceans create very costly natural borders. Very few nation states span more than one continent. Only very large and resource-rich states are able to sustain the costs of governance across oceans for longer periods of time.
- Rivers: some political borders have been formalized along natural borders formed by rivers. Some examples are; the Rio Grande border (Mexico-USA), the Rhine border (France-Germany), and the Mekong border (Thailand-Laos)

- Lakes: larger lakes create natural borders. One example is the natural border created by Lake Tanganyika (Congo-Burundi-Tanzania-Zambia)
- Forests: denser jungles or forests can create strong natural borders. One example of a natural forest border is the Amazon rain forest (Colombia-Venezuela-Guyana-Brazil-Bolivia-Peru)
- Mountain ranges: research on borders suggests that mountains have especially strong effects as natural borders. Many nations in Europe and Asia have had their political borders defined along mountain ranges.

Throughout history, technological advances have reduced the costs of transport and communication across these natural borders. This has reduced the significance of natural borders over time. As a result, political borders that have been formalized more recently — such as those in Africa or Americas — typically conform less to natural borders than very old borders — such as those in Europe or Asia — do. States whose borders conform to natural borders are, for similar reasons, more likely to be strong nation-states.

Border Economics

The presence of borders often fosters certain economic features or anomalies. Wherever two jurisdictions come into contact, special economic opportunities arise for border trade. Smuggling provides a classic case; contrariwise, a border region may flourish on the provision of excise or of import–export services — legal or quasi-legal, corrupt or corruption-free. Different regulations on either side of a border may encourage services to position themselves at or near that border: thus the provision of pornography, of prostitution, of alcohol and/or of narcotics may cluster around borders, city limits, county lines, ports and airports. In a more planned and official context, Special Economic Zones (SEZs) often tend to cluster near borders or ports.

Human economic traffic across borders (apart from kidnapping), may involve mass commuting between workplaces and residential settlements. The removal of internal barriers to commerce, as in France after the French Revolution or in Europe since the 1940s, de-emphasises border-based economic activity and fosters free trade. Euroregions are similar official structures built around commuting across borders.

Border Politics

Political borders have a variety of meanings for those whom they affect. Many borders in the world have checkpoints where border control agents inspect those crossing the boundary.

In much of Europe, such controls were abolished by the Schengen Agreement and subsequent European Union legislation.

Since the Treaty of Amsterdam, the competence to pass laws on crossing internal and external boders within the European Union and the associated Schengen States (Iceland, Norway, Switzerland, and Liechtenstein) lies exclusively within the jurisdiction of the European Union, except where states have used a specific right to opt-out (United Kingdom and Ireland, which maintain a common travel area amongst themselves).

The United States has notably increased measures taken in border control on the Canada–United States border and the United States–Mexico border during its War on Terrorism. One American writer has said that the 3600-km (2000-mile) US-Mexico border is probably "the world's longest boundary between a First World and Third World country."

Historic borders such as the Great Wall of China, the Maginot Line, and Hadrian's Wall have played a great many roles and been marked in different ways. While the stone walls, the Great Wall of China and the Roman Hadrian's Wall in Britain had military functions, the entirety of the Roman borders were very porous, a policy which encouraged Roman economic activity with its neighbors.

On the other hand, a border like the Maginot Line was entirely military and was meant to prevent any access in what was to be World War II to France by its neighbor, Germany. Germany ended up going around the Maginot Line through Belgium just as it had done in World War I.

Cross-border Regions

Macro-regional integration initiatives, such as the European Union and NAFTA, have spurred the establishment of cross-border regions. These are initiatives driven by local or regional authorities, aimed at dealing with local border-transcending problems such as transport and environmental degradation. Many cross-border regions are also active in encouraging intercultural communication and dialogue as well as cross-border economic devélopment strategies.

In Europe, the European Union provides financial support to cross-border regions via its Interreg programme. The Council of Europe has issued the Outline Convention on Transfrontier Co-operation, providing a legal framework for cross-border co-operation even though it is in practice rarely used by Euroregions.

Spatial Data

spatial data also known as *geospatial data* or *geographic information* it is the data or information that identifies the geographic location of features and boundaries on Earth, such as natural or constructed features, oceans, and more. Spatial data is usually stored as coordinates and topology, and is data that can be mapped. Spatial data is often accessed, manipulated or analyzed through Geographic Information Systems (GIS).

Types of Spatial Data

There are two types of spatial data. The geometry data type supports planar, or Euclidean (flat-earth), data. The geometry data type conforms to the Open Geospatial Consortium (OGC) Simple Features for SQL Specification version 1.1.0.

In addition, SQL Server supports the geography data type, which stores ellipsoidal (round-earth) data, such as GPS latitude and longitude coordinates.

The geometry and geography Data Types support eleven spatial data objects, or instance types. However, only seven of these instance types are instantiable; you can create and work with these instances (or instantiate them) in a database. These instances derive certain properties from their parent data types that distinguish them as Points, LineStrings, Polygons, or as multiple geometry or geography instances in a GeometryCollection.

The figure below depicts the geometry hierarchy upon which the geometry and geography data types are based. The instantiable types of geometry and geography are indicated in

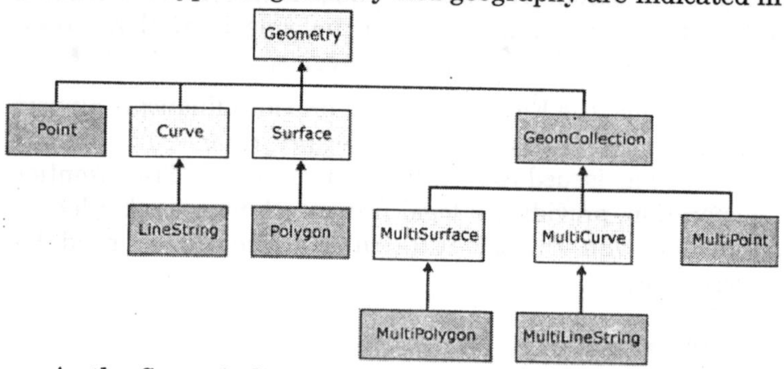

As the figure indicates, the seven instantiable types of the geometry and geography data types are Point, MultiPoint, LineString, MultiLineString, Polygon, MultiPolygon, and GeometryCollection. The geometry and geography types can recognize a specific instance as long as it is a well-formed instance, even if the instance is not defined explicitly. For example, if you define a Point instance explicitly using the STPointFromText() method, geometry and geography recognize the instance as a Point, as long as the method input is well-formed. If you define the same instance using the STGeomFromText() method, both the geometry and geography data types recognize the instance as a Point.

Differences Between the Two Data Types

The two types of spatial data often behave quite similarly, but there are some key differences in how the data is stored and manipulated.

How Connecting Edges are Defined

The defining data for LineString and Polygon types are vertices only. The connecting edge between two vertices in a geometry type is a straight line. However, the connecting edge between two vertices in a geography type is a short great elliptic arc between the two vertices. A great ellipse is the intersection of the ellipsoid with a plane through its center and a great elliptic arc is an arc segment on the great ellipse.

Measurements in Spatial Data Types

In the planar, or flat-earth, system, measurements of distances and areas are given in the same unit of measurement as coordinates. Using the geometry data type, the distance between (2, 2) and (5, 6) is 5 units, regardless of the units used.

In the ellipsoidal, or round-earth system, coordinates are given in degrees of latitude and longitude. However, lengths and areas are usually measured in meters and square meters, though the measurement may depend on the spatial reference identifier (SRID) of the geography instance. The most common unit of measurement for the geography data type is meters.

Spatial Data Infrastructure

A spatial data infrastructure (SDI) is a framework of spatial data, metadata, users and tools that are interactively connected in order to use spatial data in an efficient and flexible way.

Another definition is *the technology, policies, standards, human resources, and related activities necessary to acquire, process, distribute, use, maintain, and preserve spatial data.*

A further definition is given in Kuhn (2005) :

An SDI is a coordinated series of agreements on technology standards, institutional arrangements, and policies that enable the discovery and use of geospatial information by users and for purposes other than those it was created for.

General

Some of the main principles are that data and metadata should not be managed centrally, but by the data originator and/or owner, and that tools and services connect via computer networks to the various sources. A GIS is often the platform for deploying an individual node within an SDI. To achieve these objectives, good coordination between all the actors is necessary and the definition of standards is very important.

Due to its nature (size, cost, number of interactors) an SDI is usually government-related. An example of an existing SDI is the National Spatial Data Infrastructure (NSDI) in the United States.

At the European side, INSPIRE is a European Commission initiative to build a European SDI beyond national boundaries and ultimately the United Nations Spatial Data Infrastructure UNSDI will do the same for over 30 UN Funds, Programmes, Specialized Agencies and member countries.

Software Components

A SDI should enable the discovery and delivery of spatial data from a data repository, via a spatial service provider, to a user. As mentioned earlier it is often wished that the data provider is able to update spatial data stored in a repository. Hence, the basic software components of an SDI are :

1. a software client-to display, query, and analyse spatial data (this could be a browser or a Desktop GIS),
2. a catalogue service-for the discovery, browsing, and querying of metadata or spatial services, spatial datasets and other resources,
3. a spatial data service-allowing the delivery of the data via the Internet,
4. processing services-such as datum and projection transformations,
5. a (spatial) data repository-to store data, e.g. a Spatial database,
6. GIS software (client or desktop)-to create and update spatial data.

Besides these software components, a range of (international) technical standards are necessary that allow interaction between the different software components.

Among those are geospatial standards defined by the Open Geospatial Consortium (e.g. OGC WMS, WFS, GML etc.) and ISO (e.g. ISO 19115) for the delivery of maps, vector and raster data, but also data format and internet transfer standards by W3c consortium.

Spatial Database

A spatial database is a database that is optimized to store and query data related to objects in space, including points, lines and polygons. While typical databases can understand various numeric and character types of data, additional functionality needs to be added for databases to process spatial data types. These are typically called *geometry* or *feature*. The Open Geospatial Consortium created the Simple Features specification and sets standards for adding spatial functionality to database systems.

Features of Spatial Databases

Database systems use indexes to quickly look up values and the way that most databases index data is not optimal for spatial queries. Instead, spatial databases use a spatial index to speed up database operations.

In addition to typical SQL queries such as SELECT statements, spatial databases can perform a wide variety of spatial operations. The following query types and many more are supported by the Open Geospatial Consortium:

- Spatial Measurements: Finds the distance between points, polygon area, etc.
- Spatial Functions: Modify existing features to create new ones, for example by providing a buffer around them, intersecting features, etc.
- Spatial Predicates: Allows true/false queries such as 'is there a residence located within a mile of the area we are planning to build the landfill?'
- Constructor Functions: Creates new features with an SQL query specifying the vertices (points of nodes) which can make up lines. If the first and last vertex of a line are identical the feature can also be of the type polygon (a closed line).
- Observer Functions: Queries which return specific information about a feature such as the location of the center of a circle

Not all spatial databases support these query types.

Spatial Database Systems

- All OpenGIS Specifications compliant products [1]
- Open source spatial databases and APIs, some of which are OpenGIS compliant [2]
- Boeing's Spatial Query Server (Official Site) spatially enables Sybase ASE.
- Smallworld VMDS, the native GE Smallworld GIS database
- IBM DB2 Spatial Extender can be used to enable any edition of DB2, including the free DB2 Express-C, with support for spatial types
- Oracle Spatial
- Microsoft SQL Server has support for spatial types since version 2008
- PostgreSQL DBMS (database management system) uses the spatial extension PostGIS to implement the standardized datatype *geometry* and corresponding functions.
- MySQL DBMS implements the datatype *geometry* plus some spatial functions that haven't been implemented according to the OpenGIS specifications. Functions that test spatial relationships are limited to working with minimum bounding rectangles rather than the actual geometries. MySQL versions earlier than 5.0.16 only supported spatial data in MyISAM tables. As of MySQL 5.0.16, InnoDB, NDB, BDB, and ARCHIVE also support spatial features.
- Neo4j-Graph database that can build 1D and 2D indexes as Btree, Quadtree and Hilbert curve directly in the Graph (mathematics)
- AllegroGraph-a Graph database provides a novel mechanism for efficient storage and retrieval of two-dimensional geospatial coordinates for Resource Description Framework data. It includes an extension syntax for SPARQL queries

5

Geographic Information System

A geographic information system (GIS), or geographical information system, is any system that captures, stores, analyzes, manages, and presents data that are linked to location. In the simplest terms, GIS is the merging of cartography, statistical analysis, and database technology. GIS systems are used in cartography, remote sensing, land surveying, utility management, natural resource management, photogrammetry, geography, urban planning, emergency management, navigation, and localized search engines.

As GIS is a system, it establishes boundaries that may be jurisdictional, purpose or application oriented for which a specific GIS is developed. Hence, a GIS developed for an application, jurisdiction, enterprise, or purpose may not be necessarily interoperable or compatible with a GIS that has been developed for some other application, jurisdiction, enterprise, or purpose. What goes beyond a GIS is a spatial data infrastructure (SDI), a concept that has no such restrictive boundaries.

Therefore, in a general sense, the term describes any information system that integrates, stores, edits, analyzes, shares, and displays geographic information. In a more generic sense, GIS applications are tools that allow users to create interactive queries (user-created searches), analyze spatial information, edit data, maps, and present the results of all these operations. Geographic information science is the science underlying the geographic concepts, applications and systems, studied in degree and certificate programs at many universities.

Applications

GIS technology can be used for: earth surface based scientific investigations; resource management, reference, and projections of a geospatial nature—both manmade and natural; asset management and location planning; archaeology; environmental impact study; infrastructure assessment and development; urban planning; cartography, for a thematic and/or time based purpose; criminology; GIS data development geographic history; marketing; logistics; population and demographic studies; prospectivity mapping; location attributes applied statistical analysis; warfare assessments; and other purposes. Examples of use are: GIS may allow emergency planners to easily calculate emergency response times and the movement of response resources (for logistics) in the case of a natural disaster; GIS might be used too to find wetlands that need protection strategies regarding pollution; or GIS can be used by a company to site a new business location to take advantage of GIS data identified trends to respond to a previously under-served market. Most city and transportation systems planning offices have GIS sections.

History of Development

In 1854, John Snow depicted a cholera outbreak in London using points to represent the locations of some individual cases, possibly the earliest use of the geographic method. His study of the distribution of cholera led to the source of the disease, a contaminated water pump (the Broad Street Pump, whose handle he had disconnected, thus terminating the outbreak) within the heart of the cholera outbreak.

While the basic elements of topography and theme existed previously in cartography, the John Snow map was unique, using cartographic methods not only to depict but also to analyze clusters of geographically dependent phenomena for the first time.

The early 20th century saw the development of photolithography, by which maps were separated into layers.

Computer hardware development spurred by nuclear weapon research led to general-purpose computer "mapping" applications by the early 1960s.

The year 1962 saw the development of the world's first true operational GIS in Ottawa, Ontario, Canada by the federal Department of Forestry and Rural Development. Developed by Dr. Roger Tomlinson, it was called the "Canada Geographic Information System" (CGIS) and was used to store, analyze, and manipulate data collected for the Canada Land Inventory (CLI) – an effort to determine the land capability for rural Canada by mapping information about soils, agriculture, recreation, wildlife, waterfowl, forestry, and land use at a scale of 1:50,000. A rating classification factor was also added to permit analysis.

CGIS was the world's first such system and an improvement over "mapping" applications as it provided capabilities for overlay, measurement, and digitizing/scanning. It supported a national coordinate system that spanned the continent, coded lines as "arcs" having a true embedded topology, and it stored the attribute and locational information in separate files. As a result of this, Tomlinson has become known as the "father of GIS," particularly for his use of overlays in promoting the spatial analysis of convergent geographic data. CGIS lasted into the 1990s and built the largest digital land resource database in Canada. It was developed as a mainframe based system in support of federal and provincial resource planning and management. Its strength was continent-wide analysis of complex datasets. The CGIS was never available in a commercial form.

In 1964, Howard T Fisher formed the Laboratory for Computer Graphics and Spatial Analysis at the Harvard Graduate School of Design (LCGSA 1965-1991), where a number of important theoretical concepts in spatial data handling were developed, and which by the 1970s had distributed seminal software code and systems, such as 'SYMAP', 'GRID', and 'ODYSSEY' — which served as literal and inspirational sources for subsequent commercial development to universities, research centers, and corporations worldwide.

By the early 1980s, M&S Computing (later Intergraph), Environmental Systems Research Institute (ESRI), CARIS (Computer Aided Resource Information System) and ERDAS emerged as commercial vendors of GIS software, successfully incorporating many of the CGIS features, combining the first generation approach to separation of spatial and attribute information with a second generation approach to organizing attribute data into database structures. In parallel, the development of two public domain systems began in the late 1970s and early 1980s.

MOSS, the Map Overlay and Statistical System project started in 1977 in Fort Collins, Colorado under the auspices of the Western Energy and Land Use Team (WELUT) and the U.S. Fish and Wildlife Service. GRASS GIS was begun in 1982 by the U.S. Army Corps of Engineering Research Laboratory (USA-CERL) in Champaign, Illinois, a branch of the U.S. Army Corps of Engineers to meet the need of the U.S. military for software for land management and environmental planning. The later 1980s and 1990s industry growth were spurred on by the growing use of GIS on Unix workstations and the personal computer. By the end of the 20th century, the rapid growth in various systems had been consolidated and standardized on relatively few platforms, and users were beginning to export the concept of viewing GIS data over the Internet, requiring data format and transfer standards. More recently, a growing number of free, open source GIS packages run on a range of operating systems and can be customized to perform specific tasks.

GIS Techniques and Technology

Modern GIS technologies use digital information, for which various digitized data creation methods are used. The most common method of data creation is digitization, where a hard copy map or survey plan is transferred into a digital medium through the use of a computer-aided design (CAD) program, and geo-referencing capabilities. With the wide availability of ortho-rectified imagery (both from satellite and aerial sources), heads-up digitizing is becoming the main avenue through which geographic data is extracted. Heads-up digitizing involves the tracing of geographic data directly on top of the aerial imagery instead of by the traditional method of tracing the geographic form on a separate digitizing tablet (heads-down digitizing).

Relating Information from Different Sources

Location may be annotated by x, y, and z coordinates of longitude, latitude, and elevation, or by other geocode systems like ZIP codes or by highway mile markers. Any variable that can be located spatially can be fed into a GIS. Diverse computer databases that can be directly entered into a GIS are being produced by government agencies and nongovernment organizations. Different kinds of data in map form can be entered into or outputed from a GIS, though encoding and decoding schemas are becoming more normalized over time.

GIS Uncertainties

GIS accuracy depends upon source data, and how it is encoded to be data referenced. Land Surveyors have been able to provide a high level of positional accuracy utilizing the GPS derived positions. [Retrieved from Federal Geographic Data Committee] the high-resolution digital terrain and aerial imagery, [Retrieved NJGIN] the powerful computers, Web technology, are changing the quality, utility, and expectations of GIS to serve society on a grand scale, but nevertheless there are other source data that has an impact on the overall GIS accuracy like: paper maps that are not found to be very suitable to achieve the desired accuracy since the aging of maps affects their dimensional stability.

In developing a Digital Topographic Data Base for a GIS, topographical maps are the main source of data. Aerial photography and satellite images are extra sources for collecting data and identifying attributes which can be mapped in layers over a location facsimile of scale. The scale of a map and geographical rendering area representation type are a very important aspects since the information content depends mainly on the scale set and resulting locatability of the map's representations. In order to digitize a map, the map has to be checked within theoretical dimensions, then scanned into a raster format, and resulting raster data has to be given a theoretical dimension by a rubber sheeting/warping technology process.

Uncertainty is a significant problem in designing a GIS because spatial data tend to be used for purposes for which they were never intended. Some maps were made many decades ago, where at that time the computer industry was not even in its perspective establishments. This has led to historical reference maps without common norms. Map accuracy is a relative issue of minor importance in cartography. All maps are established for communication ends. Maps use an hsitorically constrained technology of pen and paper to communicate a view of the world to their users. Cartographers feel little need to communicate information based on accuracy, for when the same map is digitized and input into a GIS, the mode of use often changes. The new uses extend well beyond a determined domain for which the original map was intended and designed.

A quantitative analysis of maps brings accuracy issues into focus. The electronic and other equipment used to make measurements for GIS is far more precise than the machines of conventional map analysis. [Retrieved USGS]. The truth is that all geographical data are inherently inaccurate, and these inaccuracies will propagate through GIS operations in ways that are difficult to predict, yet have goals of conveyance in mind for original design. Accuracy Standards for 1:24000 Scales Map: 1:24,000 ± 40.00 feet

This means that when we see a point or attribute on a map, its "probable" location is within a +/-40 foot area of its rendered reference, according to area representations and scale.

A GIS can also convert existing digital information, which may not yet be in map form, into forms it can recognize, employ for its data analysis processes, and use in forming mapping output. For example, digital satellite images generated through remote sensing can be analyzed to produce a map-like layer of digital information about vegetative covers on land locations. Another fairly recently developed resource for naming GIS location objects is the Getty Thesaurus of Geographic Names (GTGN), which is a structured vocabulary containing about 1,000,000 names and other information about places.

Likewise, researched census or hydrological tabular data can be displayed in map-like form, serving as layers of thematic information for forming a GIS map.

Data Representation

GIS data represents real objects (such as roads, land use, elevation, trees, waterways, etc.) with digital data determining the mix. Real objects can be divided into two abstractions: discrete objects (e.g., a house) and continuous fields (such as rainfall amount, or elevations). Traditionally, there are two broad methods used to store data in a GIS for both kinds of abstractions mapping references: raster images and vector. Points, lines, and polygons are the stuff of mapped location attribute references. A new hybrid method of storing data is that of identifying point clouds, which combine three-dimensional points with RGB information at each point, returning a "3D color image". GIS Thematic maps then are becoming more and more realistically visually descriptive of what they set out to show or determine.

Raster

A raster data type is, in essence, any type of digital image represented by reducible and enlargeable grids. Anyone who is familiar with digital photography will recognize the Raster graphics pixel as the smallest individual grid unit building block of an image, hopefully not readily identified as an artifact shape until an image is produced on a very large scale. A combination of the pixels making up an image color formation scheme will compose details of an image, as is distinct from the commonly used points, lines, and polygon area location symbols of scalable vector graphics as the basis of the vector model of area attribute rendering. While a digital image is concerned with its output blending together its grid based details as an identifiable representation of reality, in a photograph or art image transferred into a computer, the raster data type will reflect a digitized abstraction of reality dealt with by grid populating tones or objects, quantities, cojoined or open boundaries, and map relief schemas. Aerial photos are one commonly used form of raster data, with one primary purpose in mind: to display a detailed image on a map area, or for the purposes of rendering its identifiable objects by digitization. Additional raster data sets used by a GIS will contain information regarding elevation, a digital elevation model, or reflectance of a particular wavelength of light, Landsat, or other electromagnetic spectrum indicators.

Raster data type consists of rows and columns of cells, with each cell storing a single value. Raster data can be images (raster images) with each pixel (or cell) containing a color value. Additional values recorded for each cell may be a discrete value, such as land use, a continuous value, such as temperature, or a null value if no data is available. While a raster cell stores a single value, it can be extended by using raster bands to represent RGB (red, green, blue) colors, colormaps (a mapping between a thematic code and RGB value), or an extended attribute table with one row for each unique cell value. The resolution of the raster data set is its cell width in ground units.

Raster data is stored in various formats; from a standard file-based structure of TIF, JPEG, etc. to binary large object (BLOB) data stored directly in a relational database management system (RDBMS) similar to other vector-based feature classes. Database storage, when properly indexed, typically allows for quicker retrieval of the raster data but can require storage of millions of significantly sized records.

Vector

In a GIS, geographical features are often expressed as vectors, by considering those features as geometrical shapes. Different geographical features are expressed by different types of geometry:

Points

In geometry, topology and related branches of mathematics a spatial point describes a specific object within a given space that consists of neither volume, area, length, nor any other higher dimensional analogue. Thus, a point is a 0-dimensional object. Because of their nature as one of the simplest geometric concepts, they are often used in one form or another as the fundamental constituents of geometry, physics, vector graphics, and many other fields.

Points in Euclidean Geometry

Points are most often considered within the framework of Euclidean geometry, where they are one of the fundamental objects. Euclid originally defined the point vaguely, as "that which has no part". In two dimensional Euclidean space, a point is represented by an ordered pair,(x, y), of numbers, where the first number conventionally represents the horizontal and is often denoted by$x$, and the second number conventionally represents the vertical and is often denoted by$y$. This idea is easily generalized to three dimensional Euclidean space, where a point is represented by an ordered triplet,(x, y, z), with the additional third number representing depth and often denoted by z. Further generalizations are represented by an ordered tuplet of n terms,$(a_1, a_2, ..., a_n)$ where n is the dimension of the space in which the point is located.

Many constructs within Euclidean geometry consist of an infinite collection of points that conform to certain axioms. This is usually represented by a set of points; As an example, a line is an infinite set of points of the form

$$L = \{(a_1, a_2, ...a_n)|a_1c_1 + a_2c_2 + ...a_nc_n = d\},$$

where $c_1$through $c_n$and $d$are constants and n is the dimension of the space. Similar constructions exist that define the plane, line segment and other related concepts.

In addition to defining points and constructs related to points, Euclid also postulated a key idea about points; he claimed that any two points can be connected by a straight line. This is easily confirmed under modern expansions of Euclidean geometry, and had lasting consequences at its introduction, allowing the construction of almost all the geometric concepts of the time. However, Euclid's postulation of points was neither complete nor definitive, as he occasionally assumed facts about points that didn't follow directly from his axioms, such as the ordering of points on the line or the existence of specific points. In spite of this, modern expansions of the system serve to remove these assumptions.

Points in Branches of Mathematics

A point in point-set topology is defined as a member of the underlying set of a topological space.

Although the notion of a point is generally considered fundamental in mainstream geometry and topology, there are some systems that forego it, e.g. noncommutative geometry and pointless topology. A "pointless space" is defined not as a set, but via some structure (algebraic or logical respectively) which looks like a well-known function space on the set: an algebra of continuous functions or an algebra of sets respectively. More precisely, such structures generalize well-known spaces of functions in a way that the operation "take a value at this point" may not be defined.

Zero-dimensional points are used for geographical features that can best be expressed by a single point reference — in other words, by simple location. Examples include wells, peaks, features of interest, and trailheads. Points convey the least amount of information of these file types. Points can also be used to represent areas when displayed at a small scale. For example, cities on a map of the world might be represented by points rather than polygons. No measurements are possible with point features.

Lines or Polylines

One-dimensional lines or polylines are used for linear features such as rivers, roads, railroads, trails, and topographic lines. Again, as with point features, linear features displayed at a small scale will be represented as linear features rather than as a polygon. Line features can measure distance.

Polygons

In geometry a polygon is traditionally a plane figure that is bounded by a closed path or *circuit*, composed of a finite sequence of straight line segments (i.e., by a closed polygonal chain). These segments are called its *edges* or *sides*, and the points where two edges meet are the polygon's *vertices* or *corners*. An *n*-gon is a polygon with *n* sides. The interior of the polygon is sometimes called its *body*. A polygon is a 2-

dimensional example of the more general polytope in any number of dimensions.

Usually two edges meeting at a corner are required to form an angle that is not straight (180°); otherwise, the line segments will be considered parts of a single edge.

The basic geometrical notion has been adapted in various ways to suit particular purposes. For example in the computer graphics (image generation) field, the term polygon has taken on a slightly altered meaning, more related to the way the shape is stored and manipulated within the computer.

Classification

Number of sides: Polygons are primarily classified by the number of sides.

Convexity: Polygons may be characterised by their degree of convexity:

- Convex: any line drawn through the polygon (and not tangent to an edge or corner) meets its boundary exactly twice.
- Non-convex: a line may be found which meets its boundary more than twice.
- Simple: the boundary of the polygon does not cross itself. All convex polygons are simple.
- Concave: Non-convex and simple.
- Star-shaped: the whole interior is visible from a single point, without crossing any edge. The polygon must be simple, and may be convex or concave.
- Self-intersecting: the boundary of the polygon crosses itself. Branko Grünbaum calls these coptic, though this term does not seem to be widely used. The term *complex* is sometimes used in contrast to *simple*, but this risks confusion with the idea of a *complex polygon* as one which exists in the complex Hilbert plane consisting of two complex dimensions.
- Star polygon: a polygon which self-intersects in a regular way.

Symmery;

- Equiangular: all its corner angles are equal.
- Cyclic: all corners lie on a single circle.
- Isogonal or vertex-transitive: all corners lie within the same symmetry orbit. The polygon is also cyclic and equiangular.
- Equilateral: all edges are of the same length. (A polygon with 5 or more sides can be *equilateral* without being *convex*.)
- Isotoxal or edge-transitive: all sides lie within the same symmetry orbit. The polygon is also equilateral.
- Regular. A polygon is regular if it is both *cyclic* and *equilateral*. A non-convex regular polygon is called a *regular star polygon*.

Miscellaneous

- Rectilinear: a polygon whose sides meet at right angles, i.e., all its interior angles are 90 or 270 degrees.
- Monotone with respect to a given line L, if every line orthogonal to L intersects the polygon not more than twice.

Properties

We will assume Euclidean geometry throughout.

Angles

Any polygon, regular or irregular, self-intersecting or simple, has as many corners as it has sides. Each corner has several angles. The two most important ones are:

- Interior angle – The sum of the interior angles of a simple n-gon is (n ” 2)ð radians or (n “ 2)180 degrees. This is because any simple n-gon can be considered to be made up of (n ” 2) triangles, each of which has an angle sum of ð radians or 180 degrees. The measure of any interior angle of a convex regular n-gon is $(1-\frac{2}{n})\pi$ radians or $180-\frac{360}{n}$ degrees. The interior angles of regular star polygons were first studied by Poinsot, in the same paper in which he describes the four regular star polyhedra.

- Exterior angle – Imagine walking around a simple *n*-gon marked on the floor. The amount you "turn" at a corner is the exterior or external angle. Walking all the way round the polygon, you make one full turn, so the sum of the exterior angles must be 360°. Moving around an n-gon in general, the sum of the exterior angles (the total amount one "turns" at the vertices) can be any integer multiple *d* of 360°, e.g. 720° for a pentagram and 0° for an angular "eight", where *d* is the density or starriness of the polygon.

The exterior angle is the supplementary angle to the interior angle. From this the sum of the interior angles can be easily confirmed, even if some interior angles are more than 180°: going clockwise around, it means that one sometime turns left instead of right, which is counted as turning a negative amount. (Thus we consider something like the winding number of the orientation of the sides, where at every vertex the contribution is between "½ and ½ winding.)

Generalizations of Polygons

In a broad sense, a polygon is an unbounded (without ends) sequence or circuit of alternating segments (sides) and angles (corners). An ordinary polygon is unbounded because the sequence closes back in itself in a loop or circuit, while an apeirogon (infinite polygon) is unbounded because it goes on for ever so you can never reach any bounding end point. The modern mathematical understanding is to describe such a structural sequence in terms of an "abstract" polygon which is a partially ordered set (poset) of elements. The interior (body) of the polygon is another element, and (for technical reasons) so is the null polytope or nullitope.

A geometric polygon is understood to be a "realization" of the associated abstract polygon; this involves some "mapping" of elements from the abstract to the geometric. Such a polygon does not have to lie in a plane, or have straight sides, or enclose an area, and individual elements can overlap or even coincide. For example a spherical polygon is drawn on the surface of a sphere, and its sides are arcs of great circles. So when we talk about "polygons" we must be careful to explain what kind we are talking about.

A digon is a closed polygon having two sides and two corners. On the sphere, we can mark two opposing points (like the North and South poles) and join them by half a great circle. Add another arc of a different great circle and you have a digon. Tile the sphere with digons and you have a polyhedron called a hosohedron. Take just one great circle instead, run it all the way round, and add just one "corner" point, and you have a monogon or henagon—although many authorities do not regard this as a proper polygon.

Other realizations of these polygons are possible on other surfaces, but in the Euclidean (flat) plane, their bodies cannot be sensibly realized and we think of them as degenerate.

The idea of a polygon has been generalized in various ways. Here is a short list of some degenerate cases (or special cases, depending on your point of view):

- Digon. Interior angle of 0° in the Euclidean plane.
- Interior angle of 180°: In the plane this gives an apeirogon, on the sphere a dihedron
- A skew polygon does not lie in a flat plane, but zigzags in three dimensions. The Petrie polygons of the regular polyhedra are classic examples.
- A spherical polygon is a circuit of sides and corners on the surface of a sphere.
- An apeirogon is an infinite sequence of sides and angles, which is not closed but it has no ends because it extends infinitely.
- A complex polygon is a figure analogous to an ordinary polygon, which exists in the complex Hilbert plane.

Naming Polygons

Individual polygons are named (and sometimes classified) according to the number of sides, combining a Greek-derived numerical prefix with the suffix-*gon*, e.g. *pentagon*, *dodecagon*. The triangle, quadrilateral or quadrangle, and nonagon are exceptions. For large numbers, mathematicians usually write the numeral itself, e.g. *17-gon*. A variable can even be used, usually *n-gon*. This is useful if the number of sides is used in a formula.

Some special polygons also have their own names; for example the regular star pentagon is also known as the pentagram.

Two-dimensional polygons are used for geographical features that cover a particular area of the earth's surface. Such features may include lakes, park boundaries, buildings, city boundaries, or land uses. Polygons convey the most amount of information of the file types. Polygon features can measure perimeter and area.

Each of these geometries is linked to a row in a database that describes their attributes. For example, a database that describes lakes may contain a lake's depth, water quality, pollution level. This information can be used to make a map to describe a particular attribute of the dataset. For example, lakes could be coloured depending on level of pollution. Different geometries can also be compared. For example, the GIS could be used to identify all wells (point geometry) that are within one kilometre of a lake (polygon geometry) that has a high level of pollution.

Vector features can be made to respect spatial integrity through the application of topology rules such as 'polygons must not overlap'. Vector data can also be used to represent continuously varying phenomena. Contour lines and triangulated irregular networks (TIN) are used to represent elevation or other continuously changing values. TINs record values at point locations, which are connected by lines to form an irregular mesh of triangles. The face of the triangles represent the terrain surface.

Advantages and Disadvantages

There are some important advantages and disadvantages to using a raster or vector data model to represent reality:

- Raster datasets record a value for all points in the area covered which may require more storage space than representing data in a vector format that can store data only where needed.
- Raster data allows easy implementation of overlay operations, which are more difficult with vector data.
- Vector data can be displayed as vector graphics used on traditional maps, whereas raster data will appear as an image that may have a blocky appearance for object boundaries. (depending on the resolution of the raster file)
- Vector data can be easier to register, scale, and re-project, which can simplify combining vector layers from different sources.
- Vector data is more compatible with relational database environments, where they can be part of a relational table as a normal column and processed using a multitude of operators.
- Vector file sizes are usually smaller than raster data, which can be 10 to 100 times larger than vector data (depending on resolution).
- Vector data is simpler to update and maintain, whereas a raster image will have to be completely reproduced. (Example: a new road is added).
- Vector data allows much more analysis capability, especially for "networks" such as roads, power, rail, telecommunications, etc. (Examples: Best route, largest port, airfields connected to two-lane highways). Raster data will not have all the characteristics of the features it displays.

Non-spatial Data

Additional non-spatial data can also be stored along with the spatial data represented by the coordinates of a vector geometry or the position of a raster cell. In vector data, the additional data contains attributes of the feature. For example, a forest inventory polygon may also have an identifier value and information about tree species. In raster data the cell value can store attribute information, but it can also be used as an identifier that can relate to records in another table.

Software is currently being developed to support spatial and non-spatial decision-making, with the solutions to spatial problems being integrated with solutions to non-spatial problems. The end result with these Flexible Spatial Decision-Making Support Systems (FSDSS) is expected to be that non-experts will be able to use GIS, along with spatial criteria, and simply integrate their non-spatial criteria to view solutions to multi-criteria problems. This system is intended to assist decision-making.

Data Capture

Data capture—entering information into the system—consumes much of the time of GIS practitioners. There are a variety of methods used to enter data into a GIS where it is stored in a digital format.

Existing data printed on paper or PET film maps can be digitized or scanned to produce digital data. A digitizer produces vector data as an operator traces points, lines, and polygon boundaries from a map. Scanning a map results in raster data that could be further processed to produce vector data.

Survey data can be directly entered into a GIS from digital data collection systems on survey instruments using a technique called Coordinate Geometry (COGO). Positions from a Global Navigation Satellite System (GNSS) like Global Positioning System (GPS), another survey tool, can also be directly entered into a GIS.

Remotely sensed data also plays an important role in data collection and consist of sensors attached to a platform. Sensors include cameras, digital scanners and LIDAR, while platforms usually consist of aircraft and satellites.

The majority of digital data currently comes from photo interpretation of aerial photographs. Soft copy workstations are used to digitize features directly from stereo pairs of digital photographs. These systems allow data to be captured in two and three dimensions, with elevations measured directly from a stereo pair using principles of photogrammetry. Currently, analog aerial photos are scanned before being entered into a soft copy system, but as high quality digital cameras become cheaper this step will be skipped.

Satellite remote sensing provides another important source of spatial data. Here satellites use different sensor packages to passively measure the reflectance from parts of the electromagnetic spectrum or radio waves that were sent out from an active sensor such as radar. Remote sensing collects raster data that can be further processed using different bands to identify objects and classes of interest, such as land cover.

When data is captured, the user should consider if the data should be captured with either a relative accuracy or absolute accuracy, since this could not only influence how information will be interpreted but also the cost of data capture.

In addition to collecting and entering spatial data, attribute data is also entered into a GIS. For vector data, this includes additional information about the objects represented in the system.

After entering data into a GIS, the data usually requires editing, to remove errors, or further processing. For vector data it must be made "topologically correct" before it can be used for some advanced analysis. For example, in a road network, lines must connect with nodes at an intersection. Errors such as undershoots and overshoots must also be removed. For scanned maps, blemishes on the source map may need to be removed from the resulting raster. For example, a fleck of dirt might connect two lines that should not be connected.

Raster-to-vector Translation

Data restructuring can be performed by a GIS to convert data into different formats. For example, a GIS may be used to convert a satellite image map to a vector structure by generating lines around all cells with the same classification, while determining the cell spatial relationships, such as adjacency or inclusion.

More advanced data processing can occur with image processing, a technique developed in the late 1960s by NASA and the private sector to provide contrast enhancement, false colour rendering and a variety of other techniques including use of two dimensional Fourier transforms.

Since digital data is collected and stored in various ways, the two data sources may not be entirely compatible. So a GIS must be able to convert geographic data from one structure to another.

Projections, Coordinate Systems and Registration

A property ownership map and a soils map might show data at different scales. Map information in a GIS must be manipulated so that it registers, or fits, with information gathered from other maps. Before the digital data can be analyzed, they may have to undergo other manipulations—projection and coordinate conversions, for example—that integrate them into a GIS. The earth can be represented by various models, each of which may provide a different set of coordinates (e.g., latitude, longitude, elevation) for any given point on the Earth's surface. The simplest model is to assume the earth is a perfect sphere. As more measurements of the earth have accumulated, the models of the earth have become more sophisticated and more accurate. In fact, there are models that apply to different areas of the earth to provide increased accuracy (e.g., North American Datum, 1927-NAD27-works well in North America, but not in Europe).

Projection is a fundamental component of map making. A projection is a mathematical means of transferring information from a model of the Earth, which represents a three-dimensional curved surface, to a two-dimensional medium—paper or a computer screen. Different projections are used for different types of maps because each projection particularly suits specific uses. For example, a projection that accurately represents the shapes of the continents will distort their relative sizes.

Since much of the information in a GIS comes from existing maps, a GIS uses the processing power of the computer to transform digital information, gathered from sources with different projections and/or different coordinate systems, to a common projection and coordinate system. For images, this process is called rectification.

Spatial Analysis with GIS

Given the vast range of spatial analysis techniques that have been developed over the past half century, any summary or review can only cover the subject to a limited depth. This is a rapidly changing field, and GIS packages are increasingly including analytical tools as standard built-in facilities or as optional toolsets, add-ins or 'analysts'. In many instances such facilities are provided by the original software suppliers (commercial vendors or collaborative non commercial development teams), whilst in other cases facilities have been developed and are provided by third parties. Furthermore, many products offer software development kits (SDKs), programming languages and language support, scripting facilities and/or special interfaces for developing one's own analytical tools or variants. The website Geospatial Analysis and associated book/ebook attempt to provide a reasonably comprehensive guide to the subject. The impact of these myriad paths to perform spatial analysis create a new dimension to business intelligence termed "spatial intelligence" which, when delivered via intranet, democratizes access to operational sorts not usually privy to this type of information.

Data Modelling

It is difficult to relate wetlands maps to rainfall amounts recorded at different points such as airports, television stations, and high schools. A GIS, however, can be used to depict two- and three-dimensional characteristics of the Earth's surface, subsurface, and atmosphere from information points. For example, a GIS can quickly generate a map with isopleth or contour lines that indicate differing amounts of rainfall.

Such a map can be thought of as a rainfall contour map. Many sophisticated methods can estimate the characteristics of surfaces from a limited number of point measurements. A two-dimensional contour map created from the surface modelling of rainfall point measurements may be overlaid and analyzed with any other map in a GIS covering the same area.

Additionally, from a series of three-dimensional points, or digital elevation model, isopleth lines representing elevation contours can be generated, along with slope analysis, shaded relief, and other elevation products. Watersheds can be easily defined for any given reach, by computing all of the areas contiguous and uphill from any given point of interest. Similarly, an expected thalweg of where surface water would want to travel in intermittent and permanent streams can be computed from elevation data in the GIS.

Topological Modelling

A GIS can recognize and analyze the spatial relationships that exist within digitally stored spatial data. These topological relationships allow complex spatial modelling and analysis to be performed. Topological relationships between geometric entities traditionally include adjacency (what adjoins what), containment (what encloses what), and proximity (how close something is to something else).

Networks

If all the factories near a wetland were accidentally to release chemicals into the river at the same time, how long would it take for a damaging amount of pollutant to enter the wetland reserve? A GIS can simulate the routing of materials along a linear network. Values such as slope, speed limit, or pipe diameter can be incorporated into network modelling to represent the flow of the phenomenon more accurately. Network modelling is commonly employed in transportation planning, hydrology modelling, and infrastructure modelling.

Cartographic Modelling

The term "cartographic modelling" was (probably) coined by Dana Tomlin in his PhD dissertation and later in his book which has the term in the title. Cartographic modelling refers to a process where several thematic layers of the same area are produced, processed, and analyzed. Tomlin used raster layers, but the overlay method can be used more generally. Operations on map layers can be combined into algorithms, and eventually into simulation or optimization models.

Map Overlay

The combination of several spatial datasets (points, lines or polygons) creates a new output vector dataset, visually similar to stacking several maps of the same region. These overlays are similar to mathematical Venn diagram overlays. A union overlay combines the geographic features and attribute tables of both inputs into a single new output. An intersect overlay defines the area where both inputs overlap and retains a set of attribute fields for each. A symmetric difference overlay defines an output area that includes the total area of both inputs except for the overlapping area.

Data extraction is a GIS process similar to vector overlay, though it can be used in either vector or raster data analysis. Rather than combining the properties and features of both datasets, data extraction involves using a "clip" or "mask" to extract the features of one data set that fall within the spatial extent of another dataset.

In raster data analysis, the overlay of datasets is accomplished through a process known as "local operation on multiple rasters" or "map algebra," through a function that combines the values of each raster's matrix. This function may weigh some inputs more than others through use of an "index model" that reflects the influence of various factors upon a geographic phenomenon.

Automated Cartography

Digital cartography and GIS both encode spatial relationships in structured formal representations. GIS is used in digital cartography modelling as a (semi)automated process of making maps, so called Automated Cartography. In practice, it can be a subset of a GIS, within which it is equivalent to the stage of visualization, since in most cases not all of the GIS functionality is used. Cartographic products can be either in a digital or in a hardcopy format. Powerful analysis techniques with different data representation can produce high-quality maps within a short time period. The main problem in Automated Cartography is to use a single set of data to produce multiple products at a variety of scales, a technique known as cartographic generalization.

Cartographic Generalization

Cartographic generalization is the method whereby

information is selected and represented on a map in a way that adapts to the scale of the display medium of the map, not necessarily preserving all intricate geographical or other cartographic details. The cartographer is given license to adjust the content within their maps to create a suitable and useful map that conveys geospatial information, while striking the right balance between the map's purpose and actuality of the subject being mapped.

Well generalized maps are those that emphasize the most important map elements while still representing the world in the most faithful and recognizable way. The level of detail and importance in what is remaining on the map must outweigh the insignificance of items that were generalized, as to preserve the distinguishing characteristics of what makes the map useful and important.

Methods

Some cartographic generalization methods include the following:

Selection

Map generalization can take many forms, and is designed to reduce the complexities of the real world by strategically reducing ancillary and unnecessary details. One way that geospatial data can be reduced is through the selection process. The cartographer can select and retain certain elements that he/she deems the most necessary or appropriate. In this method, the most important elements stand out while lesser elements are left out entirely. For example, a directional map between two points may have lesser and un-traveled roadways omitted as not to confuse the map-reader. The selection of the most direct and uncomplicated route between the two points is the most important data, and the cartographer may choose to emphasize this.

Simplification

Generalization is not a process that only removes and selects data, but also a process that simplifies it as well. Simplification is a technique where shapes of retained features are altered to enhance visibility and reduce complexity. Smaller scale maps have more simplified features than larger scale maps because they simply exhibit more area. An example of simplification is to scale and remove points along an area. Doing this to a mountain would reduce the detail in and around the mountain but would ideally not detract from the map reader interpreting the feature as such a mountain.

Combination

Simplification also takes on other roles when considering the role of combination. Overall data reduction techniques can also mean that in addition to generalizing elements of particular features, features can also be combined when their separation is irrelevant to the map focus. A mountain chain may be isolated into several smaller ridges and peaks with intermittent forest in the natural environment, but shown as a contiguous chain on the map, as determined by scale. The map reader has to, again remember, that because of scale limitations combined elements are not concise depictions of natural or manmade features.

Smoothing

Smoothing is also a process that the map maker can employ to reduce the angularity of line work. Smoothing is yet another way of simplifying the map features, but involves several other characteristics of generalization that lead into feature displacement and locational shifting. The purpose of smoothing is exhibit linework in a much less complicated and a less visually jarring way. An example of smoothing would be for a jagged roadway, cut through a mountain, to be smoothed out so that the angular turns and transitions appear much more fluid and natural.

Enhancement

Enhancement is also a method that can be employed by the cartographer to illuminate specific elements that aid in map reading. As many of the aforementioned generalizing methods focus on the reduction and omission of detail, the enhancement method concentrates on the addition of detail. Enhancement can be used to describe the true character of the feature being represented and is often used by the cartographer to highlight specific details about his or her specific knowledge, that would otherwise be left out. An example includes enhancing the detail about specific river rapids so that the map reader may know the facets of traversing the most difficult sections beforehand. Enhancement can be a valuable tool in aiding the map reader to elements that carry significant weight to the map's intent.

GIS and Automated Generalization

As GIS gained prevalence in the late 20th century and the demand for producing maps automatically increased automated generalization became an important issue for National Mapping Agencies (NMAs) and other data providers. Thereby automated generalization describes the automated extraction of data (becoming then information) regarding purpose and scale. Different researchers invented conceptual models for automated generalization:

- Gruenreich model
- Brassel & Weibel model
- McMaster & Shea model.

Besides these established model, different views on automated generalization have been established. The representation-oriented view and the process-oriented view. The first view focuses on the representation of data on different scales, which is related to the field of Multi-Representation Databases (MRDB). The latter view focuses on the process of generalization.

In the context of creating databases on different scales additionally it can be distinguished between the ladder and the star-approach. The ladder-approach is a stepwise generalization, in which each derived dataset is based on the other database of the next larger scale. The star-approach describes the derived data on all scales is based on a single (large-scale) data base.

Operators in Automated Generalization

Automated generalization had always to compete with manual cartographers, therefore the manual generalization process was studied intensively. These studies resulted early in different generalization operators. By now there is no clear classification of operators available and it is doubtful if a comprehensive classification will evolve in future.

Geostatistics

Geostatistics is a point-pattern analysis that produces field predictions from data points. It is a way of looking at the statistical properties of those special data. It is different from general applications of statistics because it employs the use of graph theory and matrix algebra to reduce the number of parameters in the data. Only the second-order properties of the GIS data are analyzed.

When phenomena are measured, the observation methods dictate the accuracy of any subsequent analysis. Due to the nature of the data (e.g. traffic patterns in an urban environment; weather patterns over the Pacific Ocean), a constant or dynamic degree of precision is always lost in the measurement. This loss of precision is determined from the scale and distribution of the data collection.

To determine the statistical relevance of the analysis, an average is determined so that points (gradients) outside of any immediate measurement can be included to determine their predicted behaviour. This is due to the limitations of the applied statistic and data collection methods, and interpolation is required to predict the behaviour of particles, points, and locations that are not directly measurable.

Interpolation is the process by which a surface is created, usually a raster dataset, through the input of data collected at a number of sample points. There are several forms of interpolation, each which treats the data differently, depending on the properties of the data set. In comparing interpolation methods, the first consideration should be whether or not the source data will change (exact or approximate). Next is whether the method is subjective, a human interpretation, or objective. Then there is the nature of transitions between points: are they abrupt or gradual. Finally, there is whether a method is global (it uses the entire data set to form the model), or local where an algorithm is repeated for a small section of terrain.

Interpolation is a justified measurement because of a spatial autocorrelation principle that recognizes that data collected at any position will have a great similarity to, or influence of those locations within its immediate vicinity.

Digital elevation models (DEM), triangulated irregular networks (TIN), edge finding algorithms, Thiessen polygons, Fourier analysis, (weighted) moving averages, inverse distance weighting, kriging, spline, and trend surface analysis are all mathematical methods to produce interpolative data.

Address Geocoding

Geocoding is interpolating spatial locations (X,Y coordinates) from street addresses or any other spatially referenced data such as ZIP Codes, parcel lots and address locations. A reference theme is required to geocode individual addresses, such as a road centerline file with address ranges. The individual address locations have historically been interpolated, or estimated, by examining address ranges along a road segment. These are usually provided in the form of a table or database. The GIS will then place a dot approximately where that address belongs along the segment of centerline. For example, an address point of 500 will be at the midpoint of a line segment that starts with address 1 and ends with address 1000. Geocoding can also be applied against actual parcel data, typically from municipal tax maps. In this case, the result of the geocoding will be an actually positioned space as opposed to an interpolated point. This approach is being increasingly used to provide more precise location information. There are several potentially dangerous caveats that are often overlooked when using interpolation.

Various algorithms are used to help with address matching when the spellings of addresses differ. Address information that a particular entity or organization has data on, such as the post office, may not entirely match the reference theme. There could be variations in street name spelling, community name, etc. Consequently, the user generally has the ability to make matching criteria more stringent, or to relax those parameters so that more addresses will be mapped. Care must be taken to review the results so as not to map addresses incorrectly due to overzealous matching parameters.

Reverse Geocoding

Reverse geocoding is the process of returning an estimated street address number as it relates to a given coordinate. For example, a user can click on a road centerline theme (thus providing a coordinate) and have information returned that reflects the estimated house number. This house number is interpolated from a range assigned to that road segment. If the user clicks at the midpoint of a segment that starts with address 1 and ends with 100, the returned value will be somewhere near 50. Note that reverse geocoding does not return actual addresses, only estimates of what should be there based on the predetermined range.

Data Output and Cartography

Cartography is the design and production of maps, or visual representations of spatial data. The vast majority of modern cartography is done with the help of computers, usually using a GIS but production quality cartography is also achieved by importing layers into a design program to refine it. Most GIS software gives the user substantial control over the appearance of the data.

Cartographic work serves two major functions:

First, it produces graphics on the screen or on paper that convey the results of analysis to the people who make decisions about resources. Wall maps and other graphics can be generated, allowing the viewer to visualize and thereby understand the results of analyses or simulations of potential events. Web Map Servers facilitate distribution of generated maps through web browsers using various implementations of web-based application programming interfaces (AJAX, Java, Flash, etc.).

Second, other database information can be generated for further analysis or use. An example would be a list of all addresses within one mile (1.6 km) of a toxic spill.

Graphic Display Techniques

Traditional maps are abstractions of the real world, a sampling of important elements portrayed on a sheet of paper with symbols to represent physical objects. People who use maps must interpret these symbols. Topographic maps show the shape of land surface with contour lines or with shaded relief.

Today, graphic display techniques such as shading based on altitude in a GIS can make relationships among map elements visible, heightening one's ability to extract and analyze information. For example, two types of data were combined in a GIS to produce a perspective view of a portion of San Mateo County, California.

- The digital elevation model, consisting of surface elevations recorded on a 30-meter horizontal grid, shows high elevations as white and low elevation as black.
- The accompanying Landsat Thematic Mapper image shows a false-color infrared image looking down at the same area in 30-meter pixels, or picture elements, for the same coordinate points, pixel by pixel, as the elevation information.

A GIS was used to register and combine the two images to render the three-dimensional perspective view looking down the San Andreas Fault, using the Thematic Mapper image pixels, but shaded using the elevation of the landforms. The GIS display depends on the viewing point of the observer and time of day of the display, to properly render the shadows created by the sun's rays at that latitude, longitude, and time of day.

An archeochrome is a new way of displaying spatial data. It is a thematic on a 3D map that is applied to a specific building or a part of a building. It is suited to the visual display of heat loss data.

Spatial ETL

Spatial ETL tools provide the data processing functionality of traditional Extract, Transform, Load (ETL) software, but with a primary focus on the ability to manage spatial data. They provide GIS users with the ability to translate data between different standards and proprietary formats, whilst geometrically transforming the data en-route.

GIS Developments

Many disciplines can benefit from GIS technology. An active GIS market has resulted in lower costs and continual improvements in the hardware and software components of GIS. These developments will, in turn, result in a much wider use of the technology throughout science, government, business, and industry, with applications including real estate, public health, crime mapping, national defence, sustainable development, natural resources, landscape architecture, archaeology, regional and community planning, transportation and logistics. GIS is also diverging into location-based services (LBS). LBS allows GPS enabled mobile devices to display their location in relation to fixed assets (nearest restaurant, gas station, fire hydrant), mobile assets (friends, children, police car) or to relay their position back to a central server for display or other processing. These services continue to develop with the increased integration of GPS functionality with increasingly powerful mobile electronics (cell phones, PDAs, laptops).

OGC Standards

The Open Geospatial Consortium (OGC) is an international industry consortium of 384 companies, government agencies, universities and individuals participating in a consensus process to develop publicly available geoprocessing specifications. Open interfaces and protocols defined by OpenGIS Specifications support interoperable solutions that "geo-enable" the Web, wireless and location-based services, and mainstream IT, and empower technology developers to make complex spatial information and services accessible and useful with all kinds of applications. Open Geospatial Consortium (OGC) protocols include Web Map Service (WMS) and Web Feature Service (WFS).

GIS products are broken down by the OGC into two categories, based on how completely and accurately the software follows the OGC specifications.

Compliant Products are software products that comply to OGC's OpenGIS Specifications. When a product has been tested and certified as compliant through the OGC Testing Program, the product is automatically registered as "compliant" on this site.

Implementing Products are software products that implement OpenGIS Specifications but have not yet passed a compliance test. Compliance tests are not available for all specifications. Developers can register their products as implementing draft or approved specifications, though OGC reserves the right to review and verify each entry.

Web Mapping

Web mapping is the process of designing, implementing, generating and delivering maps on the World Wide Web and its product. While web mapping primarily deals with technological issues, web cartography additionally studies theoretic aspects: the use of web maps, the evaluation and optimization of techniques and workflows, the usability of web maps, social aspects, and more. Web GIS is similar to web mapping but with an emphasis on analysis, processing of project specific geodata and exploratory aspects. Often the terms web GIS and web mapping are used synonymously, even if they don't mean exactly the same. In fact, the border between web maps and web GIS is blurry. Web maps are often a presentation media in web GIS and web maps are increasingly gaining analytical capabilities. A special case of web maps are mobile maps, displayed on mobile computing devices, such as mobile phones, smart phones, PDAs, GPS and other devices. If the maps on these devices are displayed by a mobile web browser or web user agent, they can be regarded as mobile web maps. If the mobile web maps also display context and location sensitive information, such as points of interest, the term Location-based services is frequently used."

"The use of the web as a dissemination medium for maps can be regarded as a major advancement in cartography and opens many new opportunities, such as realtime maps, cheaper dissemination, more frequent and cheaper updates of data and software, personalized map content, distributed data sources and sharing of geographic information. It also implicates many challenges due to technical restrictions (low display resolution and limited bandwidth, in particular with mobile computing devices, many of which are physically small, and use slow wireless Internet connections), copyright and security issues, reliability issues and technical complexity. While the first web maps were primarily static, due to technical restrictions, today's web maps can be fully interactive and integrate multiple media. This means that both web mapping and web cartography also have to deal with interactivity, usability and multimedia issues."

Development and Implementation

The advent of web mapping can be regarded as a major new trend in cartography. Previously, cartography was restricted to a few companies, institutes and mapping agencies, requiring expensive and complex hardware and software as well as skilled cartographers and geomatics engineers. With web mapping, freely available mapping technologies and geodata potentially allow every skilled person to produce web maps, with expensive geodata and technical complexity (data harmonization, missing standards) being two of the remaining barriers preventing web mapping from fully going mainstream.

The cheap and easy transfer of geodata across the internet allows the integration of distributed data sources, opening opportunities that go beyond the possibilities of disjoint data storage. Everyone with minimal knowhow and infrastructure can become a geodata provider.

These facts can be regarded both as an advantage and a disadvantage. While it allows everyone to produce maps and considerably enlarges the audience, it also puts geodata in the hands of untrained people who potentially violate cartographic and geographic principles and introduce flaws during the preparation, analysis and presentation of geographic and cartographic data. Educating the general public about geographic analysis and cartographic methods and principles should therefore be a priority to the cartography community.

Types of Web Maps

A first classification of web maps has been made by Kraak. He distinguished *static* and *dynamic* web maps and further distinguished *interactive* and *view only* web maps. However, today in the light of an increased number of different web map types, this classification needs some revision. Today, there are additional possibilities regarding distributed data sources, collaborative maps, personalized maps, etc.

Analytic Web Maps

These web maps offer GIS analysis, either with geodata provided, or with geodata uploaded by the map user. As already mentioned, the borderline between analytic web maps and web GIS is blurry. Often, parts of the analysis are carried out by a serverside GIS and the client displays the result of the analysis. As web clients gain more and more capabilities, this task sharing may gradually shift.

Animated Web Maps

Animated Maps show changes in the map over time by animating one of the graphical or temporal variables. Various data and multimedia formats and technologies allow the display of animated web maps: SVG, Adobe Flash, Java, Quicktime, etc., also with varying degrees of interaction. Examples for animated web maps are weather maps, maps displaying dynamic natural or other phenomena (such as water currents, wind patterns, traffic flow, trade flow, communication patterns,social studies projects, and for college life, etc.).

Collaborative Web Maps

Collaborative maps are still new, immature and complex to implement, but show a lot of potential. The method parallels the Wikipedia project where various people collaborate to create and improve maps on the web. Technically, an application allowing simultaneous editing across the web would have to ensure that geometric features being edited by one person are locked, so they can't be edited by other persons at the same time. Also, a minimal quality check would have to be made, before data goes public. Some collaborative map projects:

- OpenStreetMap
- WikiMapia
- meta:Maps-survey of Wikimedia map proposals on Wikipedia:Meta

Customisable Web Maps

Web maps in this category are usually more complex web mapping systems that offer APIs for reuse in other people's web pages and products. Example for such a system with an API for reuse are the Open Layers Framework, Yahoo! Maps and Google Maps.

Distributed Web Maps

These are maps created from a distributed data source. The WMS protocol offers a standardised method to access maps on other servers.

WMS servers can collect these different sources, reproject the map layers, if necessary, and send them back as a combined image containing all requested map layers. One server may offer a topographic base map, while other servers may offer thematic layers.

Dynamically Created Web Maps

These maps are created on demand each time the user reloads the webpages, often from dynamic data sources, such as databases. The webserver generates the map using a web map server or a self written software. Some applications refer to depictions as hyper maps. One of the example is-Bhoosampada] by Indian Space Research Organizations.

Hyper Maps

Any approach offering the planar presentation of a portion of an n-dimensional orthogonal web map structure with the option to chose the axes for depiction from the dimensions.

Interactive Web Maps

Interactivity is one of the major advantages of screen based maps and web maps. It helps to compensate for the disadvantages of screen and web maps. Interactivity helps to explore maps, change map parameters, navigate and interact with the map, reveal additional information, link to other resources, and much more. Technically, it is achieved through the combination of events, scripting and DOM manipulations.

Online Atlases

Atlas projects often went through a renaissance when they made a transition to a web based project. In the past, atlas projects often suffered from expensive map production, small circulation and limited audience. Updates were expensive to produce and took a long time until they hit the public. Many atlas projects, after moving to the web, can now reach a wider audience, produce cheaper, provide a larger number of maps and map types and integrate with and benefit from other web resources. Some atlases even ceased their printed editions after going online, sometimes offering printing on demand features from the online edition. Some atlases (primarily from North America) also offer raw data downloads of the underlying geospatial data sources.

Personalized Web Maps

Personalized web maps allow the map user to apply his own data filtering, selective content and the application of personal styling and map symbolization. The OGC (Open Geospatial Consortium) provides the SLD standard (Styled Layer Description) that may be sent to a WMS server for the application of individual styles. This implies that the content and data structure of the remote WMS server is properly documented.

Realtime Web Maps

Realtime maps show the situation of a phenomenon in close to realtime (only a few seconds or minutes delay). Data is collected by sensors and the maps are generated or updated at regular intervals or immediately on demand. Examples are weather maps, traffic maps or vehicle monitoring systems.

Static Web Maps

Static web pages are *view only* with no animation and interactivity. They are only created once, often manually and infrequently updated. Typical graphics formats for static web maps are PNG, JPEG, GIF, or TIFF (e.g., drg) for raster files, SVG, PDF or SWF for vector files. Often, these maps are scanned paper maps and had not been designed as screen maps. Paper maps have a much higher resolution and information density than typical computer displays of the same physical size, and might be unreadable when displayed on screens at the wrong resolution.

Temporal Web Maps

Any depiction of a portion of an n-dimensional orthogonal web map structure in a planar projection with time as one of the coordinate axes.

Advantages of Web Maps

- Web maps can easily *deliver up to date information.* If maps are generated automatically from databases, they can display information in almost realtime. They don't need to be printed, mastered and distributed. Examples:
  * A map displaying election results, as soon as the election results become available.
  * A map displaying the traffic situation near realtime by using traffic data collected by sensor networks.
  * A map showing the current locations of mass transit vehicles such as buses or trains, allowing patrons to minimize their waiting time at stops or stations, or be aware of delays in service.
  * Weather maps, such as NEXRAD.
- *Software and hardware infrastructure for web maps is cheap.* Web server hardware is cheaply available and many open source tools exist for producing web maps.
- *Product updates can easily be distributed.* Because web maps distribute both logic and data with each request or loading, product updates can happen every time the web user reloads the application. In traditional cartography, when dealing with printed maps or interactive maps distributed on offline media (CD, DVD, etc.), a map update caused serious efforts, triggering a reprint or remastering as well as a redistribution of the media. With web maps, data and product updates are easier, cheaper, and faster, and can occur more often.

- *They work across browsers and operating systems.* If web maps are implemented based on open standards, the underlying operating system and browser do not matter.
- *Web maps can combine distributed data sources.* Using open standards and documented APIs one can integrate (*mash up*) different data sources, if the projection system, map scale and data quality match. The use of centralized data sources removes the burden for individual organizations to maintain copies of the same data sets. The down side is that one has to rely on and trust the external data sources.
- *Web maps allow for personalization.* By using user profiles, personal filters and personal styling and symbolization, users can configure and design their own maps, if the web mapping systems supports personalization. Accessibility issues can be treated in the same way. If users can store their favourite colors and patterns they can avoid color combinations they can't easily distinguish (e.g. due to color blindness).
- *Web maps enable collaborative mapping.* Similar to the Wikipedia project, web mapping technologies, such as DHTML/Ajax, SVG, Java, Adobe Flash, etc. enable distributed data acquisition and collaborative efforts. Examples for such projects are the OpenStreetMap project or the Google Earth community. As with other open projects, quality assurance is very important, however!
- *Web maps support hyperlinking to other information on the web.* Just like any other web page or a wiki, web maps can act like an index to other information on the web. Any sensitive area in a map, a label text, etc. can provide hyperlinks to additional information. As an example a map showing public transport options can directly link to the corresponding section in the online train time table.
- *It is easy to integrate multimedia in and with web maps.* Current web browsers support the playback of

video, audio and animation (SVG, SWF, Quicktime, and other multimedia frameworks).

Disadvantages of Web Maps and Problematic Issues

- *Reliability issues* – the reliability of the internet and web server infrastructure is not yet good enough. Especially if a web map relies on external, distributed data sources, the original author often cannot guarantee the availability of the information.
- *Geodata is expensive* – Unlike in the US, where geodata collected by governmental institutions is usually available for free or cheap, geodata is usually very expensive in Europe or other parts of the world.
- *Bandwidth issues* – Web maps usually need a relatively high bandwidth.
- *Limited screen space* – Like with other screen based maps, web maps have the problem of limited screen space. This is in particular a problem for mobile web maps and location based services where maps have to be displayed in very small screens with resolutions as low as 100×100 pixels. Hopefully, technological advances will help to overcome these limitations.
- *Quality and accuracy issues* – Many web maps are of poor quality, both in symbolization, content and data accuracy.
- *Complex to develop* – Despite the increasing availability of free and commercial tools to create web mapping and web GIS applications, it is still a complex task to create interactive web maps. Many technologies, modules, services and data sources have to be mastered and integrated.
- *Immature development tools* – Compared to the development of standalone applications with integrated development tools, the development and debugging environments of a conglomerate of different web technologies is still awkward and uncomfortable.
- *Copyright issues* – Many people are still reluctant to publish geodata, especially in the light that geodata is

expensive in some parts of the world. They fear copyright infringements of other people using their data without proper requests for permission.

- *Privacy issues* – With detailed information available and the combination of distributed data sources, it is possible to find out and combine a lot of private and personal information of individual persons. Properties and estates of individuals are now accessible through high resolution aerial and satellite images throughout the world to anyone.

History of Web Mapping

This section contains some of the milestones of web mapping, online mapping services and atlases. Because web mapping depends on enabling technologies of the web, this section also includes a few milestones of the web.

- 1989-09: *Birth of the WWW*, WWW invented at CERN for the exchange of research documents.
- 1990-12: *First Web Browser and Web Server*, Tim Berners-Lee wrote first web browser and web server.
- 1991-04: HTTP 0.9 protocol, Initial design of the HTTP protocol for communication between browser and server.
- 1991-06: *ViolaWWW 0.8 Browser*, The first popular web browser. Written for X11 on Unix.
- 1991-08: *WWW project announced in public newsgroup*, This is regarded as the debut date of the Web. Announced in newsgroup alt.hypertext.
- 1992-06: HTTP 1.0 protocol, Version 1.0 of the HTTP protocol. Introduces the POST method and persistent connections.
- 1993-04: *CERN announced web as free*, CERN announced that access to the web will be free for all. The web gained critical mass.
- 1993-06: HTML 1.0. The first version of HTML, published by T. Berners-Lee and Dan Connolly.
- 1993-07: *Xerox PARC Map Viewer*, The first mapserver based on CGI/Perl, allowed reprojection styling and definition of map extent.

- 1994-06: *The National Atlas of Canada*, The first version of the National Atlas of Canada was released. Can be regarded as the first online atlas.
- 1994-10: *Netscape Browser 0.9 (Mosaic)*, The first version of the highly popular browser Netscape Navigator.
- 1995-03: *Java 1.0*, The first public version of Java.
- 1995-11: HTML 2.0, Introduced forms, file upload, internationalization and client-side image maps.
- 1995-12: *Javascript 1.0*, Introduced first script based interactivity.
- 1995: *MapGuide*, First introduced as Argus MapGuide.
- 1996-01: *JDK 1.0*, First version of the Sun JDK.
- 1996-02: *Mapquest*, The first popular online Address Matching and Routing Service with mapping output.
- 1996-06: *MultiMap*, The UK-based MultiMap website launched offering online mapping, routing and location based services. Grew into one of the most popular UK web sites.
- 1996-11: Geomedia WebMap 1.0, First version of Geomedia WebMap, already supports vector graphics through the use of ActiveCGM.
- 1996-fall: *MapGuide*, Autodesk acquired Argus Technologies.and introduced Autodesk MapGuide 2.0.
- 1996-12: *Macromedia Flash 1.0*, First version of the Macromedia Flash plugin.
- 1997-01: HTML 3.2, Introduced tables, applets, script elements, multimedia elements, flowtext around images, etc.
- 1997-03:Norwegian company Mapnet launches application for www.epi.no with active POI layer for real estate listings.
- 1997-06: *US Online National Atlas Initiative*, The USGS received the mandate to coordinate and create the online National Atlas of the United States of America [2].
- 1997-07: UMN MapServer 1.0, Developed as Part of the

NASA ForNet Project. Grew out of the need to deliver remote sensing data across the web for foresters.

- 1997-12: HTML 4.0, Introduced styling with CSS, absolute and relative positioning of elements, frames, object element, etc.
- 1998-06: *Terraserver USA*, A Web Map Service serving aerial images (mainly b+w) and USGS DRGs was released. One of the first popular WMS. This service is a joint effort of USGS, Microsoft and HP.
- 1998-07: UMN MapServer 2.0, Added reprojection support (PROJ.4).
- 1998-08: MapObjects Internet Map Server, ESRI's entry into the web mapping business.
- 1999-03: HTTP 1.1 protocol, Version 1.1 of the HTTP protocol. Introduces the request pipelining for multiple connections between server and client. This version is still in use as of 2007.
- 1999-08: *National Atlas of Canada, 6th edition*, This new version was launched at the ICA 1999 conference in Ottawa. Introduced many new features and topics. Is being improved gradually, since then, and kept up-to-date with technical advancements.
- 2000-02: ArcIMS 3.0, The first public release of ESRI's ArcIMS.
- 2000-06: ESRI Geography Network, ESRI founded Geography Network to distribute data and web map services.
- 2000-06: UMN MapServer 3.0, Developed as part of the NASA TerraSIP Project. This is also the first public, open source release of UMN Mapserver. Added raster support and support for TrueType fonts (FreeType).
- 2000-08: *Flash Player 5*, This introduced ActionScript 1.0 (ECMAScript compatible).
- 2001-06: MapScript [3] 1.0 for UMN MapServer, Adds a lot of flexibility to UMN MapServer solutions.
- 2001-09: SVG 1.0 W3C Recommendation, SVG (Scalable Vector Graphics) 1.0 became a W3C Recommendation.

- 2001-09: *Tirolatlas*, A highly interactive online atlas, the first to be based on the SVG standard.
- 2002-06: UMN MapServer 3.5, Added support for PostGIS and ArcSDE. Version 3.6 adds initial OGC WMS support.
- 2002-07: ArcIMS 4.0, Version 4 of the ArcIMS web map server.
- 2003-01: SVG 1.1 W3C Recommendation, SVG 1.1 became a W3C Recommendation. This introduced the mobile profiles SVG Tiny and SVG Basic.
- 2003-06: *NASA World Wind*, NASA World Wind Released. An open virtual globe that loads data from distributed resources across the internet. Terrain and buildings can be viewed 3 dimensionally. The (XML based) markup language allows users to integrate their own personal content. This virtual globe needs special software and doesn't run in a web browser.
- 2003-07: UMN MapServer 4.0, Adds 24bit raster output support and support for PDF and SWF.
- 2003-09: *Flash Player 7*, This introduced ActionScript 2.0 (ECMAScript 2.0 compatible (improved object orientation)). Also initial Video Playback support.
- 2004-07: OpenStreetMap was founded by Steve Coast. OSM is a web based collaborative project to create a world map under a free license.
- 2005-01: Nikolas Schiller creates the interactive "Inaugural Map" of downtown Washington, DC
- 2005-02: *Google Maps*, The first version of Google Maps. Based on raster tiles organized in a quad tree scheme, data loading done with XMLHttpRequests. This mapping application became highly popular on the web, also because it allowed other people to integrate google map services into their own website.
- 2005-04: UMN MapServer 4.6, Adds support for SVG.
- 2005-06: *Google Earth*, The first version of Google Earth was released building on the virtual globe metaphor. Terrain and buildings can be viewed 3 dimensionally.

The KML (XML based) markup language allows users to integrate their own personal content. This virtual globe needs special software and doesn't run in a web browser.

- 2005-11: *Firefox 1.5*, First Firefox release with native SVG support. Supports Scripting but no animation.
- 2006-05: *Wikimapia* Launched
- 2006-06: *Opera 9*, Opera releases version 9 with extensive SVG support (including scripting and animation).
- 2006-08: SVG 1.2 Mobile Candidate Recommendation, This SVG Mobile Profile introduces improved multimedia support and many features required to build online Rich Internet Applications.
- 2009-01 Nokia makes Ovi Maps free on its smartphones.

Web Mapping Technologies

The potential number of technologies to implement web mapping projects is almost infinite. Any programming environment, programming language and serverside framework can be used to implement web mapping projects. In any case, both server and client side technologies have to be used. Following is a list of potential and popular server and client side technologies utilized for web mapping.

Server Side Technologies

- Web server – The webserver is responsible for handling http requests by web browsers and other user agents. In the simplest case they serve static files, such as HTML pages or static image files. Web servers also handle authentication, content negotiation, server side includes, URL rewriting and forward requests to dynamic resources, such as CGI applications or serverside scripting languages. The functionality of a webserver can usually be enhanced using modules or extensions. The most popular web server is Apache, followed by Microsoft Internet Information Server and others.

* CGI (common gateway interface) applications are executables running on the webserver under the environment and user permissions of the webserver user. They may be written in any programming language (compiled) or scripting language (e.g. perl). A CGI application implements the common gateway interface protocol, processes the information sent by the client, does whatever the application should do and sends the result back in a web-readable form to the client. As an example a web browser may send a request to a CGI application for getting a web map with a certain map extent, styling and map layer combination. The result is an image format, e.g. JPEG, PNG or SVG. For performance enhancements one can also install CGI applications such as FastCGI. This loads the application after the web server is started and keeps the application in memory, eliminating the need to spawn a separate process each time a request is being made.
* Alternatively, one can use scripting languages built into the webserver as a module, such as PHP, Perl, Python, ASP, Ruby, etc. If built into the web server as a module, the scripting engine is already loaded and doesn't have to be loaded each time a request is being made.

- Web application servers are middleware which connects various software components with the web server and a programming language. As an example, a web application server can enable the communication between the API of a GIS and the webserver, a spatial database or other proprietary applications. Typical web application servers are written in Java, C, C++, C# or other scripting languages. Web application servers are also useful when developing complex realtime web mapping applications or Web GIS.
- Spatial databases are usually object relational databases enhanced with geographic data types, methods and properties. They are necessary whenever a web mapping application has to deal with dynamic data (that changes

frequently) or with huge amount of geographic data. Spatial databases allow spatial queries, sub selects, reprojections, geometry manipulations and offer various import and export formats. A popular example for an open source spatial database is PostGIS. MySQL also implements some spatial features, although not as mature as PostGIS. Commercial alternatives are Oracle Spatial or spatial extensions of Microsoft SQL Server and IBM DB2. The OGC Simple Features for SQL Specification is a standard geometry data model and operator set for spatial databases. Most spatial databases implement this OGC standard.

- WMS server are specialized web mapping servers implemented as a CGI application, Java Servlet or other web application server. They either work as a standalone web server or in collaboration with existing web servers or web application servers (the general case). WMS Servers can generate maps on request, using parameters, such as map layer order, styling/ symbolization, map extent, data format, projection, etc. The OGC Consortium defined the WMS standard to define the map requests and return data formats. Typical image formats for the map result are PNG, JPEG, GIF or SVG. There are open source WMS Servers such as UMN Mapserver and Mapnik. Commercial alternatives exist from most commercial GIS vendors, such as ESRI ArcIMS, ArcGIS Server, GeoClip, Intergraph Geomedia WebMap, and others.

Client Side Technologies

- Web browser – In the simplest setup, only a web browser is required. All modern web browsers support the display of HTML and raster images (JPEG, PNG and GIF format). Some solutions require additional plugins.
  * ECMAScript support – ECMAScript is the standardized version of JavaScript. It is necessary to implement client side interaction, refactoring of the DOM of a webpage and for doing network requests. ECMAScript is currently part of any modern web browser.

* Events support – Various events are necessary to implement interactive client side maps. Events can trigger script execution or SMIL operations. We distinguish between:
    a Mouse events (mousedown, mouseup, mouseover, mousemove, click)
    b Keyboard events (keydown, keypress, keyup)
    c State events (load, unload, abort, error)
    d Mutation events (reacts on modifications of the DOM tree, e.g. DOMNodeInserted)
    e SMIL animation events (reacts on different states in SMIL animation, beginEvent, endEvent, repeatEvent)
    f UI events (focusin, focusout, activate)
    g SVG specific events (SVGZoom, SVGScroll, SVGResize)
* Network requests – This is necessary to load additional data and content into a web page. Most modern browsers provide the XMLHttpRequest object which allows for get and post http requests and provides some feedback on the data loading state. The data received can be processed by ECMAScript and can be included into the current DOM tree of the web page/web map. SVG user agents alternatively provide the getURL() and postURL() methods for network requests. It is recommended to test for the existence of a network request method and provide alternatives if one method isn't present. As an example, a wrapper function could handle the network requests and test whether XMLHttpRequests or getURL() or alternative methods are available and choose the best one available. These network requests are also known under the term Ajax.
* DOM support – The Document Object Model provides a language independent API for the manipulation of the document tree of the webpage.

It exposes properties of the individual nodes of the document tree, allows to insert new nodes, delete nodes, reorder nodes and change existing nodes. DOM support is included in any modern web browser. DOM support together with scripting is also known as DHTML or Dynamic HTML. Google Maps and many other web mapping sites use a combination of DHTML, Ajax, SVG and VML.

* SVG support or SVG image support – SVG is the abbreviation of "Scalable Vector Graphics" and integrates vector graphics, raster graphics and text. SVG also supports animation, internationalization, interactivity, scripting and XML based extension mechanisms. SVG is a huge step forward when it comes to delivering high quality, interactive maps. At the time of writing (2007–01), SVG is natively supported in Mozilla/Firefox >version 1.5, Opera >version 9 and the developer version of Safari/ Webkit. Internet Explorer users still need the Adobe SVG viewer plugin provided by Adobe.
* Java support – some browsers still provide old versions of the Java virtual machine. An alternative is the use of the Sun Java Plugin. Java is a full featured programming language that can be used to create very sophisticated and interactive web maps. The Java2D and Java3D libraries provide 2d and 3d vector graphics support. The creation of Java based web maps requires a lot of programming know how. Adrian Herzog discusses the use of Java applets for the presentation of interactive choroplethe and cartogram maps.
* Web browser plugins
    a Adobe Acrobat – provides vector graphics and high quality printing support. Allows toggling of map layers, hyper links, multimedia embedding, some basic interactivity and scripting (ECMAScript).
    b Adobe Flash – provides vector graphics, animation and multimedia support. Allows

the creation of sophisticated interactive maps, as with Java and SVG. Features a programming language (ActionScript) which is similar to ECMAScript. Supports Audio and Video.

c Apple Quicktime – Adds support for additional image formats, video, audio and Quicktime VR (Panorama Images). Only available to Mac OS X and Windows.

d Adobe SVG viewer – provide SVG 1.0 support for web browsers, only required for Internet Explorer Users, because it doesn't yet natively support SVG. The Adobe SVG viewer isn't developed any further and only fills the gap until Internet Explorer gains native SVG support.

e Sun Java plugin provides support for newer and advanced Java Features.

Global Change, Climate History Program and Prediction of its Impact

Maps have traditionally been used to explore the Earth and to exploit its resources. GIS technology, as an expansion of cartographic science, has enhanced the efficiency and analytic power of traditional mapping. Now, as the scientific community recognizes the environmental consequences of anthropogenic activities influencing climate change, GIS technology is becoming an essential tool to understand the impacts of this change over time. GIS enables the combination of various sources of data with existing maps and up-to-date information from earth observation satellites along with the outputs of climate change models. This can help in understanding the effects of climate change on complex natural systems. One of the classic examples of this is the study of Arctic Ice Melting. But this is only a theoretical computer model of a theoretical scenario.

The outputs from a GIS in the form of maps combined with satellite imagery allow researchers to view their subjects in ways that literally never have been seen before. The images are also invaluable for conveying the effects of climate change to non-scientists.

Prediction of the impact of climate change inherently involves many uncertainties stemming from data and models. GIS incorporated with uncertainty theory has been used to model the coastal impact of climate change, including inundation due to sea-level rise and storm erosion.

Adding the Dimension of Time

The condition of the Earth's surface, atmosphere, and subsurface can be examined by feeding satellite data into a GIS. GIS technology gives researchers the ability to examine the variations in Earth processes over days, months, and years.

As an example, the changes in vegetation vigor through a growing season can be animated to determine when drought was most extensive in a particular region. The resulting graphic, known as a normalized vegetation index, represents a rough measure of plant health. Working with two variables over time would then allow researchers to detect regional differences in the lag between a decline in rainfall and its effect on vegetation.

GIS technology and the availability of digital data on regional and global scales enable such analyses. The satellite sensor output used to generate a vegetation graphic is produced for example by the Advanced Very High Resolution Radiometer (AVHRR).

This sensor system detects the amounts of energy reflected from the Earth's surface across various bands of the spectrum for surface areas of about 1 square kilometer. The satellite sensor produces images of a particular location on the Earth twice a day. AVHRR and more recently the Moderate-Resolution Imaging Spectroradiometer (MODIS) are only two of many sensor systems used for Earth surface analysis. More sensors will follow, generating ever greater amounts of data.

GIS and related technology will help greatly in the management and analysis of these large volumes of data, allowing for better understanding of terrestrial processes and better management of human activities to maintain world economic vitality and environmental quality.

In addition to the integration of time in environmental studies, GIS is also being explored for its ability to track and model the progress of humans throughout their daily routines. A concrete example of progress in this area is the recent release of time-specific population data by the US Census. In this data set, the populations of cities are shown for daytime and evening hours highlighting the pattern of concentration and dispersion generated by North American commuting patterns. The manipulation and generation of data required to produce this data would not have been possible without GIS.

Using models to project the data held by a GIS forward in time have enabled planners to test policy decisions. These systems are known as Spatial Decision Support Systems.

Semantics

Tools and technologies emerging from the W3C's Semantic Web Activity are proving useful for data integration problems in information systems. Correspondingly, such technologies have been proposed as a means to facilitate interoperability and data reuse among GIS applications and also to enable new analysis mechanisms.

Ontologies are a key component of this semantic approach as they allow a formal, machine-readable specification of the concepts and relationships in a given domain. This in turn allows a GIS to focus on the intended meaning of data rather than its syntax or structure. For example, reasoning that a land cover type classified as *deciduous needleleaf trees* in one dataset is a specialization of land cover type *forest* in another more roughly classified dataset can help a GIS automatically merge the two datasets under the more general land cover classification. Tentative ontologies have been developed in areas related to GIS applications, for example the hydrology ontology developed by the Ordnance Survey in the United Kingdom and the SWEET ontologies developed by NASA's Jet Propulsion Laboratory. Also, simpler ontologies and semantic metadata standards are being proposed by the W3C Geo Incubator Group to represent geospatial data on the web.

Recent research results in this area can be seen in the International Conference on Geospatial Semantics and the Terra

Cognita — Directions to the Geospatial Semantic Web workshop at the International Semantic Web Conference.

Society

With the popularization of GIS in decision making, scholars have begun to scrutinize the social implications of GIS. It has been argued that the production, distribution, utilization, and representation of geographic information are largely related with the social context. Other related topics include discussion on copyright, privacy, and censorship. A more optimistic social approach to GIS adoption is to use it as a tool for public participation.

The Role of Geographic Information Systems

- GIS technology, data structures and analytical techniques are gradually being incorporated into a wide range of management and decision-making operations
- numerous examples of applications of GIS are available in many different journals and are frequent topics of presentations at conferences in the natural and social sciences
- in order to understand the range of applicability of GIS it is necessary to characterize the multitude of applications in some logical way so that similarities and differences between approaches and needs can be examined
- an understanding of this range of needs is critical for those who will be dealing with the procurement and management of a GIS

Functional Classification

One way to classify GIS applications is by functional characteristics of the systems.

This would include a consideration of:

1. characteristics of the data such as:
   * themes
   * precision required
   * data model

2. GIS functions
   * which of the range of possible GIS functions does the application rely on?
     a e.g. address matching, overlay?
3. products
   * e.g. does the application support queries, one-time video maps and/or hardcopy maps?

Aclassification based on these characteristics quickly becomes fuzzy since GIS is a flexible tool whose great strength is the ability to integrate data themes, functionality and output.

GIS as a decision support tool ;

- another way to classify GIS is by the kinds of decisions that are supported by the GIS
- several definitions of GIS identify its role in decision-making
- decision support is an excellent goal for GIS, however:
  * decisions range from major (which foreign aid project to support with limited budget?) to minor (which way to turn at next intersection?)
  * difficult to know when GIS was used to make decisions except in cases of major decisions
- decision support is a good basis for definition of GIS, but not for differentiating between applications since individual GIS systems are generally used to make several different kinds of decisions

Core groups of GIS activity ;

- GIS field is a loose coalescence of groups of users, managers, academics and professionals all working with spatial information
- each group has a distinct educational and "cultural" background
  * each has associated societies, magazines and journals, conferences, traditions
  * as a result, each identifies itself with particular ways of approaching particular sets of problems

- interactions occur between groups through joint memberships, joint conferences, umbrella organizations
- these groups or cultures, then, are another basis for characterizing application areas
- the core groups of GIS activity can be seen to be comprised of:

Mature technologies which interact with GIS, sharing its technology and creating data for it

* surveying and engineering
* cartography
* remote sensing

Management and decision-making groups

* resource inventory and management
* urban planning (Urban Information Systems)
* land records for taxation and ownership control (Land Information Systems)
* facilities management (AM/FM)
* marketing and retail planning
* vehicle routing and scheduling

Science and research activities at universities and government labs

- this and the next 5 units (Units 52-56) examine each of these groups of GIS activity seeking to find distinctions and similarities between them
- begin in this unit with a quick review of the relationship between the mature technologies and GIS and finish with a look at the role of GIS in science
- there are two areas of GIS application in cartography: 1. automation of the map-making process 2. production of new forms of maps resulting from analysis, manipulation of data
    * the second is closer to the concept of GIS although both use similar technology.

Computers in Cartography

- first efforts to automate the map-making process occurred in early 1960s
- major advantage of automation is in ease of editing
  - objects can be moved around digital map without redrafting
  - scale and projection change are relatively easy
- differences between automated mapping and GIS are frequently emphasized
  - mapping requires: knowledge of positions of objects, limited number of attributes
  - GIS requires: knowledge of positions of objects, attributes, relationships between objects
  - hence distinction between "cartographic" and "topological" databases
- "analytical" cartography involves analysis of mapped data
  - has much in common with some aspects of GIS analysis
- cartography plays a vital role in the success of GIS
  - supplies principles of design of map output products-how to make them easy to read and interpret?
  - represents centuries of development of expertise in compiling, handling, displaying geographical data
- widespread feeling that conversion to digital technology:
  - is inevitable
  - will revolutionize the field through new techniques.

Organizations

- both professional and academic organizations in most countries
  - International Cartographic Association (ICA)
- well-developed training and education programs, journals, continuing research.

Adoption

- now is some use of digital technology in almost all aspects of the map production process
- the term "desktop mapping" emphasizes the accessibility of one form of automated cartography in the same way that page formatting programs have led to the success of "desktop publishing"

Surveying and Engineering

- surveying is concerned with the measurement of locations of objects on the Earth's surface, particularly property boundaries
    * all 3 dimensions are important-vertical as well as horizontal positions
    * accuracy below 0.1 m is necessary
- the locations of a limited number of sites are fixed extremely accurately through precision instruments and measurements
    * these sites are monuments or benchmarks-the geodetic control network
    * this is the function of geodesy or geodetic science
- using these accurate benchmarks for reference, large numbers of locations can then be accurately determined relative to the fixed monuments
- surveying is an important supplier of data to GIS
    * however, it is not directly concerned with role of GIS as a decision-making tool
- some civil engineers now use GIS technology, especially digital elevation models and associated functionality, to assist in planning construction
    * e.g. to make calculations of quantities of earth to be moved in construction projects such as building highways
    * e.g. to visualize the effects of major construction projects such as dams.

Recent advances in technology ;

- instruments:
  - * locations captured by measuring device in digital form, downloaded to database-the "total station"
  - * new GPS (global positioning system) instruments determine location from satellites, supplementing the geodetic control network
- direct linkage of surveying instruments to spatial databases
  - * thus suppliers of surveying equipment have entered the GIS field as vendors.

Characteristics of application area ;

- scale:
  - * large-surveying often accurate to mm
  - * engineering calculations require high DEM resolution
- data model:
  - * survey data is exclusively vector
- lineage:
  - * for legal reasons the source of survey data is important
    - a e.g. instruments, benchmarks used, name of surveyor, date
  - * most systems do not yet allow such lineage information to be stored directly with the data.

Organizations ;

- surveying and engineering are mature professional fields based on scientific methods, with organizations, conferences, courses, journals, systems of accreditation
- introduction of GIS technology has not radically altered the profession.

Remote Sensing

- like surveying, is a data producing field

- acquires knowledge about the Earth's surface from airborne or space platforms
- elaborate, well-developed technology and techniques
  - * instruments for data capture-high spatial and spectral resolution
  - * transmission of data, processing, archiving
  - * interpreting and classifying images
- two major roles for GIS concepts:
  - * quality and value of product is enhanced by use of additional ("ancillary") data to improve accuracy of classification
    - a e.g. knowledge of ground elevation from a DEM allows shadows to be removed from images
  - * to be useful in decision-making, product needs to be combined with other layers less readily observed from space
    - a e.g. political boundaries
- remote sensing continues to be an active research area
  - * new instruments need to be evaluated for applications in different fields
  - * careful research is needed to realize the enormous potential of the technology
  - * volume of accumulated data is increasing rapidly.

Characteristics of application area ;

- scale:
  - * a full range of spatial resolutions, depending on altitude, characteristics of instrument
- data model:
  - * data is captured exclusively in raster form (pixels)
  - * classified images may be converted to vector form for output, or for input to GIS systems
- interfacing with GIS is a current development direction
  - * both areas have developed extensive software systems

* in remote sensing, systems include image processing functionality
* interfacing is not difficult technically-however, there may be substantial incompatibilities in data models, format standards and spatial resolution
* many GIS vendors include functions to convert data from remote sensing systems and to display vector data on satellite image backdrops
* true integration of vector GIS and raster image processing systems is not yet available.

Organizations ;

- because of continuing emphasis on research, there is heavy representation from government and academic research
- the growth curve of remote sensing occurred about a decade earlier than GIS.

Science and Research

- growing interest in using GIS technology to support scientific research
  * to support investigations of global environment-global science
  * to search for factors causing patterns of disease-epidemiology
  * to understand changes in patterns of settlement, distributions of population groups within cities-anthropology, demography, social geography
  * to understand relationships between species distribution and habitats-landscape ecology
- GIS has been called an enabling technology for science because of the breadth of potential uses as a tool
- Ron Abler (Pennsylvania State University) has compared GIS to tools like microscopes, Xerox machines, telescopes in its potential for support of research.

Analogy to statistical packages ;

- major statistical packages-SAS, SPSS, BMD, S etc.-developed over past 20 years

  * primarily developed to apply statistical tools in scientific research
  * subsequent applications in consulting, business
  * recent introduction of graphics, mapping capabilities for display of results, e.g. SAS/GRAPH
- unlike statistical packages, GIS development has been driven by applications other than scientific research
- lack of tools for spatial analysis has meant that the role of location in explaining phenomena has been difficult to evaluate
  * locational information has been available in map libraries but hard to interface with other information, not part of digital research environment
- potential for GIS to play an important role in scientific research
  * GIS supports spatial analysis as statistical packages support statistical analysis.

Characteristics of application area ;

- scale:
  * very large (archaeology) to very small (global science)
- functionality:
  * overlay to combine, correlate different variables
  * ability to interface GIS with complex modelling packages, statistical packages
  * interpolation
  * visualization of data
  * potential for 3D, time-dependent applications.

Organizations ;

- no forum for exclusive discussion of role of GIS in science (similar problems in statistics)
  * particularly in the non-technical fields in the social sciences

- discussion confined to individual disciplines
- geography is the only discipline with a general concern for spatial analysis and supporting tools
    * however, in most US universities geography is a small, relatively weak and unknown discipline
    * in other countries, (e.g. UK) geography is recognized as a strong traditional discipline, with distinguished roots in social and physical science research.

6

Data Mapping in Geography

Data mapping is the process of creating data element mappings between two distinct data models. Data mapping is used as a first step for a wide variety of data integration tasks including:

- Data transformation or data mediation between a data source and a destination
- Identification of data relationships as part of data lineage analysis
- Discovery of hidden sensitive data such as the last four digits social security number hidden in another user id as part of a data masking or de-identification project
- Consolidation of multiple databases into a single data base and identifying redundant columns of data for consolidation or elimination.

For example, a company that would like to transmit and receive purchases and invoices with other companies might use data mapping to create data maps from a company's data to standardized ANSI ASC X12 messages for items such as purchase orders and invoices.

Standards

X12 standards are generic Electronic Data Interchange (EDI) standards designed to allow a company to exchange data with any other company, regardless of industry. The standards are maintained by the Accredited Standards Committee X12

(ASC X12), with the American National Standards Institute (ANSI) accredited to set standards for EDI. The X12 standards are often called ANSI ASC X12 standards.

In the future, tools based on semantic web languages such as Resource Description Framework (RDF), the Web Ontology Language (OWL) and standardized metadata registry will make data mapping a more automatic process. This process will be accelerated if each application performed metadata publishing. Full automated data mapping is a very difficult problem.

Hand-coded, Graphical Manual

Data mappings can be done in a variety of ways using procedural code, creating XSLT transforms or by using graphical mapping tools that automatically generate executable transformation programs. These are graphical tools that allow a user to "draw" lines from fields in one set of data to fields in another. Some graphical data mapping tools allow users to "Auto-connect" a source and a destination. This feature is dependent on the source and destination data element name being the same. Transformation programs are automatically created in SQL, XSLT, Java programming language or C++. These kinds of graphical tools are found in most ETL Tools (Extract, Transform, Load Tools) as the primary means of entering data maps to support data movement.

Data-driven Mapping

This is the newest approach in data mapping and involves simultaneously evaluating actual data values in two data sources using heuristics and statistics to automatically discover complex mappings between two data sets. This approach is used to find transformations between two data sets and will discover substrings, concatenations, arithmetic, case statements as well as other kinds of transformation logic. This approach also discovers data exceptions that do not follow the discovered transformation logic.

Semantic Mapping

Semantic mapping is similar to the auto-connect feature of data mappers with the exception that a metadata registry can be consulted to look up data element synonyms. For example, if the source system lists FirstName but the destination lists PersonGivenName, the mappings will still be made if these data elements are listed as synonyms in the metadata registry. Semantic mapping is only able to discover exact matches between columns of data and will not discover any transformation logic or exceptions between columns.

Visual Analytics

Visual analytics is an outgrowth of the fields information visualization and scientific visualization, that focuses on analytical reasoning facilitated by interactive visual interfaces.

Overview

Visual analytics is "the science of analytical reasoning facilitated by visual interactive interfaces.". It can attack certain problems whose size, complexity, and need for closely coupled human and machine analysis may make them otherwise intractable. Visual analytics advances science and technology developments in analytical reasoning, interaction, data transformations and representations for computation and visualization, analytic reporting, and technology transition. As a research agenda, visual analytics brings together several scientific and technical communities from computer science, information visualization, cognitive and perceptual sciences, interactive design, graphic design, and social sciences.

Visual analytics integrates new computational and theory-based tools with innovative interactive techniques and visual representations to enable human-information discourse. The design of the tools and techniques is based on cognitive, design, and perceptual principles. This science of analytical reasoning provides the reasoning framework upon which one can build both strategic and tactical visual analytics technologies for threat analysis, prevention, and response. Analytical reasoning is central to the analyst's task of applying human judgments to reach conclusions from a combination of evidence and assumptions.

Visual analytics has some overlapping goals and techniques with information visualization and scientific visualization. There is currently no clear consensus on the boundaries between these fields, but broadly speaking the three areas can be distinguished as follows.

Scientific visualization deals with data that has a natural geometric structure (e.g., MRI data, wind flows). Information visualization handles abstract data structures such as trees or graphs. Visual analytics is especially concerned with sensemaking and reasoning.

Visual analytics seeks to marry techniques from information visualization with techniques from computational transformation and analysis of data. Information visualization itself forms part of the direct interface between user and machine. Information visualization amplifies human cognitive capabilities in six basic ways:

1. by increasing cognitive resources, such as by using a visual resource to expand human working memory,
2. by reducing search, such as by representing a large amount of data in a small space,
3. by enhancing the recognition of patterns, such as when information is organized in space by its time relationships,
4. by supporting the easy perceptual inference of relationships that are otherwise more difficult to induce,
5. by perceptual monitoring of a large number of potential events, and
6. by providing a manipulable medium that, unlike static diagrams, enables the exploration of a space of parameter values.

These capabilities of information visualization, combined with computational data analysis, can be applied to analytic reasoning to support the sense-making process.

Topics

Scope

Visual analytics is a multidisciplinary field that includes the following focus areas:

- Analytical reasoning techniques that enable users to obtain deep insights that directly support assessment, planning, and decision making
- Data representations and transformations that convert all types of conflicting and dynamic data in ways that support visualization and analysis
- Techniques to support production, presentation, and dissemination of the results of an analysis to communicate information in the appropriate context to a variety of audiences.
- Visual representations and interaction techniques that take advantage of the human eye's broad bandwidth pathway into the mind to allow users to see, explore, and understand large amounts of information at once

Analytical Reasoning Techniques

Analytical reasoning techniques are the method by which users obtain deep insights that directly support situation assessment, planning, and decision making. Visual analytics must facilitate high-quality human judgment with a limited investment of the analysts' time. Visual analytics tools must enable diverse analytical tasks such as:

- Understanding past and present situations quickly, as well as the trends and events that have produced current conditions
- Identifying possible alternative futures and their warning signs
- Monitoring current events for emergence of warning signs as well as unexpected events
- Determining indicators of the intent of an action or an individual
- Supporting the decision maker in times of crisis.

These tasks will be conducted through a combination of individual and collaborative analysis, often under extreme time pressure. Visual analytics must enable hypothesis-based and scenario-based analytical techniques, providing support for the analyst to reason based on the available evidence.

Data Representations

Data representations are structured forms suitable for computer-based transformations. These structures must exist in the original data or be derivable from the data themselves. They must retain the information and knowledge content and the related context within the original data to the greatest degree possible. The structures of underlying data representations are generally neither accessible nor intuitive to the user of the visual analytics tool. They are frequently more complex in nature than the original data and are not necessarily smaller in size than the original data. The structures of the data representations may contain hundreds or thousands of dimensions and be unintelligible to a person, but they must be transformable into lower-dimensional representations for visualization and analysis.

Theories of Visualization

Theories of visualization are:

- "Semiology of Graphics" in 1967 written by Jacques Bertin e
- "Languages of Art" from 1977 by Nelson Goodman
- Jock D. Mackinlay's "Automated design of optimal visualization" (APT) from 1986, and
- Leland Wilkinson's "Grammar of Graphics" from 1998.

Visual Representations

Visual representations translate data into a visible form that highlights important features, including commonalities and anomalies. These visual representations make it easy for users to perceive salient aspects of their data quickly. Augmenting the cognitive reasoning process with perceptual reasoning through visual representations permits the analytical reasoning process to become faster and more focused.

Process

The input for the data sets used in the visual analytics process are heterogeneous data sources (i.e., the internet, newspapers, books, scientific experiments, expert systems).

From these rich sources, the data sets $S = S_1, ..., S_m$ are chosen, whereas each S_i, *i* " *(1,..., m)* consists of attributes $A_{i1}, ..., A_{ik}$. The goal or output of the process is insight *I*. Insight is either directly obtained from the set of created visualizations *V* or through confirmation of hypotheses *H* as the results of automated analysis methods. This formalization of the visual analytics process is illustrated in the following figure. Arrows represent the transitions from one set to another one.

More formal the visual analytics process is a transformation $F : S \rightarrow I$, whereas *F* is a concatenation of functions *f* – $\{D_W, V_X, H_Y, U_Z\}$ defined as follows:

D_W describes the basic data pre-processing functionality with $D_W : S \rightarrow S$ *and* W – $\{T, C, SL, I\}$ including data transformation functions D_T, data cleaning functions D_C, data selection functions D_{SL} and data integration functions D_I that are needed to make analysis functions applicable to the data set.

V_W, W – $\{S, H\}$ symbolizes the visualization functions, which are either functions visualizing data $V_S : S$ '! V or functions visualizing hypotheses $V_H : H \rightarrow V$.

H_Y, Y – $\{S, V\}$ represents the hypotheses generation process. We distinguish between functions that generate hyphotheses from data $H_S : S \rightarrow H$ and functions that generate hypotheses from visualizations $H_V : V \rightarrow H$.

Moreover, user interactions U_Z, Z – $\{V, H, CV, CH\}$ are an integral part of the visual analytics process. User interactions can either effect only visualizations $U_V : V \rightarrow V$ (i.e., selecting or zooming), or can effect only hypotheses $U_H : H \rightarrow H$ by generating a new hypotheses from given ones. Furthermore, insight can be concluded from visualizations $U_{CV} : V \rightarrow I$ or from hypotheses $U_{CH} : H \rightarrow I$.

The typical data pre-processing applying data cleaning, data integration and data transformation functions is defined as $D_P = D_T(D_I(D_C(S_1, ..., S_n)))$. After the pre-processing step either automated analysis methods $H_S = \{f_{s1}, ..., f_{sq}\}$ (i.e., statistics, data mining, etc.) or visualization methods $V_S : S \rightarrow V$, $V_S = \{f_{v1}, ..., f_{vs}\}$ are applied to the data, in order to reveal patterns as shown in the figure above.

An application: Intelligent Multi-Agent System for Knowledge discovery. Researchers are working on the design and development of systems that enhance human-information interaction in information analysis and discovery for diverse applications, such as intelligence analysis and bio-informatics.

Argument Mapping

An argument map is a visual representation of the structure of an argument in informal logic. It includes the components of an argument such as a main contention, premises, co-premises, objections, rebuttals and lemmas. Typically an argument map is a "box and arrow" diagram with boxes corresponding to propositions and arrows corresponding to relationships such as evidential support. Argument mapping is often designed to support deliberation over issues, ideas and arguments in Wicked problems.

Argument Maps are often used in the teaching of reasoning and critical thinking, and can support the analysis of pros and cons when deliberating over wicked problems.

Business Decision Mapping

Business Decision Mapping (BDM) is a technique for making decisions, particularly the kind of decisions that often need to be made in business. It involves using diagrams to help articulate and work through the decision problem, from initial recognition of the need through to communication of the decision and the thinking behind it.

BDM is designed for use in making deliberative decisions-those made based on canvassing and weighing up the arguments. It is also qualitative-although numbers may be involved, the main considerations are qualitatively specified and there is no calculation-based route to the right decision. In these two key elements, BDM is similar to the natural or typical way of making decisions.

However, it differs from typical, informal decision making by providing a structured, semi-formal framework, and using visual language, taking advantage of our ability to grasp and make sense of information faster and more easily when it is graphically presented.

BDM is centred on the creation of a decision map-a single diagram that brings together in one organized structure all the fundamental elements of a decision, and that functions as a focus of collaboration.

BDM aims to support the decision process, making it easier, more reliable and more accountable. It addresses some major problems that can afflict business decision making the way it is generally done, including stress, anxiety, time pressure, lost thinking and inefficiency. By mapping the decision problem, the options, the arguments and all relevant evidence visually using BDM, the decision maker can avoid holding a large amount of information in his or her head, is able to make a more complete and transparent analysis and can generate a record of the thinking behind the final decision.

Cartography

In cartography, technology has continually changed in order to meet the demands of new generations of mapmakers and map users. The first maps were manually constructed with brushes and parchment; therefore, varied in quality and were limited in distribution. The advent of magnetic devices, such as the compass and much later, magnetic storage devices, allowed for the creation of far more accurate maps and the ability to store and manipulate them digitally.

Advances in mechanical devices such as the printing press, quadrant and vernier, allowed for the mass production of maps and the ability to make accurate reproductions from more accurate data. Optical technology, such as the telescope, sextant and other devices that use telescopes, allowed for accurate surveying of land and the ability of mapmakers and navigators to find their latitude by measuring angles to the North Star at night or the sun at noon.

Advances in photochemical technology, such as the lithographic and photochemical processes, have allowed for the creation of maps that have fine details, do not distort in shape and resist moisture and wear. This also eliminated the need for engraving, which further shortened the time it takes to make and reproduce maps. Advances in electronic technology in the 20th century ushered in another revolution in cartography. Ready availability of computers and peripherals such as monitors, plotters, printers, scanners (remote and document) and analytic stereo plotters, along with computer programs for visualization, image processing, spatial analysis, and database management, have democratized and greatly expanded the making of maps. The ability to superimpose spatially located variables onto existing maps created new uses for maps and new industries to explore and exploit these potentials.

These days most commercial-quality maps are made using software that falls into one of three main types: CAD, GIS and specialized illustration software. Spatial information can be stored in a database, from which it can be extracted on demand. These tools lead to increasingly dynamic, interactive maps that can be manipulated digitally.

Map Types

General vs Thematic Cartography

In understanding basic maps, the field of cartography can be divided into two general categories: general cartography and thematic cartography. General cartography involves those maps that are constructed for a general audience and thus contain a variety of features. General maps exhibit many reference and location systems and often are produced in a series. For example, the 1:24,000 scale topographic maps of the United States Geological Survey (USGS) are a standard as compared to the 1:50,000 scale Canadian maps. The government of the UK produces the classic 1:50,000 (replacing the older 1 inch to 1 mile) "Ordnance Survey" maps of the entire UK and with a range of correlated larger-and smaller-scale maps of great detail.

Thematic cartography involves maps of specific geographic themes, oriented toward specific audiences. A couple of examples might be a dot map showing corn production in Indiana or a shaded area map of Ohio counties, divided into numerical choropleth classes. As the volume of geographic data has exploded over the last century, thematic cartography has become increasingly useful and necessary to interpret spatial, cultural and social data. An orienteering map combines both general and thematic cartography, designed for a very specific user community. The most prominent thematic element is shading, that indicates degrees of difficulty of travel due to vegetation. The vegetation itself is not identified, merely classified by the difficulty ("fight") that it presents.

Topographic vs Topological

A topographic map is primarily concerned with the topographic description of a place, including (especially in the 20th century) the use of contour lines showing elevation. Terrain or relief can be shown in a variety of ways.

A topological map is a very general type of map, the kind you might sketch on a napkin. It often disregards scale and detail in the interest of clarity of communicating specific route or relational information. Beck's London Underground map is an iconic example. Though the most widely used map of "The Tube," it preserves little of reality: It varies scale constantly and abruptly, it straightens curved tracks, and it contorts directions haphazardly. The only topography on it is the River Thames, letting the reader know whether a station is north or south of the river. That and the topology of station order and interchanges between train lines are all that is left of the geographic space. Yet those are all a typical passenger wishes to know, so the map fulfills its purpose.

Map Design

Map Projection

A map projection is any method of representing the surface of a sphere or other shape on a plane. Map projections are necessary for creating maps. All map projections distort the surface in some fashion. Depending on the purpose of the map, some distortions are acceptable and others are not; therefore different map projections exist in order to preserve some properties of the sphere-like body at the expense of other properties. There is no limit to the number of possible map projections.

Background

For simplicity, this article usually assumes that the surface to be mapped is the surface of a sphere. However, the Earth and other sufficiently large celestial bodies are generally better modeled as oblate spheroids, and small objects such as asteroids may have irregular shapes. These other surfaces can be mapped as well. Therefore, more generally, a map projection is any method of "flattening" into a plane a continuous surface having curvature in all three spatial dimensions.

Projection as used here is not limited to perspective projections, such as those resulting from casting a shadow on a screen. Rather, any mathematical function transforming coordinates from the curved surface to the plane is a projection.

Carl Friedrich Gauss's Theorema Egregium proved that a sphere cannot be represented on a plane without distortion. Since any method of representing a sphere's surface on a plane is a map projection, all map projections distort. Every distinct map projection distorts in a distinct way. The study of map projections is the characterization of these distortions.

A map of the earth is a representation of a curved surface on a plane. Therefore a map projection must have been used to create the map, and, conversely, maps could not exist without map projections. Maps can be more useful than globes in many situations: they are more compact and easier to store; they readily accommodate an enormous range of scales; they are viewed easily on computer displays; they can facilitate measuring properties of the terrain being mapped; they can show larger portions of the Earth's surface at once; and they are cheaper to produce and transport. These useful traits of maps motivate the development of map projections.

Metric Properties of Maps

Many properties can be measured on the Earth's surface independently of its geography. Some of these properties are:

- Area
- Shape
- Direction

- Bearing
- Distance
- Scale.

Map projections can be constructed to preserve one or more of these properties, though not all of them simultaneously. Each projection preserves or compromises or approximates basic metric properties in different ways. The purpose of the map determines which projection should form the base for the map. Because many purposes exist for maps, many projections have been created to suit those purposes.

Another major concern that drives the choice of a projection is the compatibility of data sets. Data sets are geographic information. As such, their collection depends on the chosen model of the Earth.

Different models assign slightly different coordinates to the same location, so it is important that the model be known and that the chosen projection be compatible with that model. On small areas (large scale) data compatibility issues are more important since metric distortions are minimal at this level. In very large areas (small scale), on the other hand, distortion is a more important factor to consider.

Which Map is Best?

Cartographers have long despaired over publishers' inapt use of the Mercator. As a 1943 New York Times editorial states, "*...The time has come to discard [the Mercator] for something that represents the continents and directions less deceptively... Although its usage... has diminished... it is still highly popular as a wall map apparently in part because, as a rectangular map, it fills a rectangular wall space with more map, and clearly because its familiarity breeds more popularity.*"

The Peters map controversy motivated the American Cartographic Association (now Cartography and Geographic Information Society) to produce a series of booklets (including *Which Map is Best*) designed to educate the public about map projections and distortion in maps. In 1989 and 1990, after some internal debate, seven North American geographic organizations adopted the following resolution, which rejected all rectangular world maps, a category that includes both the Mercator and the Gall–Peters projections:

WHEREAS, the earth is round with a coordinate system composed entirely of circles, and

WHEREAS, flat world maps are more useful than globe maps, but flattening the globe surface necessarily greatly changes the appearance of Earth's features and coordinate systems, and

WHEREAS, world maps have a powerful and lasting effect on people's impressions of the shapes and sizes of lands and seas, their arrangement, and the nature of the coordinate system, and

WHEREAS, frequently seeing a greatly distorted map tends to make it "look right,"

THEREFORE, we strongly urge book and map publishers, the media and government agencies to cease using rectangular world maps for general purposes or artistic displays. Such maps promote serious, erroneous conceptions by severely distorting large sections of the world, by showing the round Earth as having straight edges and sharp corners, by representing most distances and direct routes incorrectly, and by portraying the circular coordinate system as a squared grid. The most widely displayed rectangular world map is the Mercator (in fact a navigational diagram devised for nautical charts), but other rectangular world maps proposed as replacements for the Mercator also display a greatly distorted image of the spherical Earth.

Construction of a Map Projection

The creation of a map projection involves three steps:

1. Selection of a model for the shape of the Earth or planetary body (usually choosing between a sphere or ellipsoid). Because the Earth's actual shape is irregular, information is lost in this step.
2. Transformation of geographic coordinates (longitude and latitude) to Cartesian (x,y) or polar plane coordinates. Cartesian coordinates normally have a

simple relation to eastings and northings defined on a grid superimposed on the projection.

Some of the simplest map projections are literally projections, as obtained by placing a light source at some definite point relative to the globe and projecting its features onto a specified surface. This is not the case for most projections which are defined only in terms of mathematical formulae that have no direct physical interpretation.

Choosing a Projection Surface

A surface that can be unfolded or unrolled into a plane or sheet without stretching, tearing or shrinking is called a *developable surface.* The cylinder, cone and of course the plane are all developable surfaces. The sphere and ellipsoid are not developable surfaces. As noted in the introduction, any projection of a sphere (or an ellipsoid) onto a plane will have to distort the image. (To compare, one cannot flatten an orange peel without tearing or warping it.)

One way of describing a projection is first to project from the Earth's surface to a developable surface such as a cylinder or cone, and then to unroll the surface into a plane. While the first step inevitably distorts some properties of the globe, the developable surface can then be unfolded without further distortion.

Aspects of the Projection

Once a choice is made between projecting onto a cylinder, cone, or plane, the aspect of the shape must be specified. The aspect describes how the developable surface is placed relative to the globe: it may be *normal* (such that the surface's axis of symmetry coincides with the Earth's axis), *transverse* (at right angles to the Earth's axis) or *oblique* (any angle in between). The developable surface may also be either *tangent* or *secant* to the sphere or ellipsoid. Tangent means the surface touches but does not slice through the globe; secant means the surface does slice through the globe. Insofar as preserving metric properties goes, it is never advantageous to move the developable surface away from contact with the globe, so that possibility is not discussed here.

Scale

A globe is the only way to represent the earth with constant scale throughout the entire map in all directions. A map cannot achieve that property for any area, no matter how small. It can, however, achieve constant scale along specific lines.

Some possible properties are:

- The scale depends on location, but not on direction. This is equivalent to preservation of angles, the defining characteristic of a conformal map.
- Scale is constant along any parallel in the direction of the parallel. This applies for any cylindrical or pseudocylindrical projection in normal aspect.
- Combination of the above: the scale depends on latitude only, not on longitude or direction. This applies for the Mercator projection in normal aspect.
- Scale is constant along all straight lines radiating from a particular geographic location. This is the defining characteristic of an equidistant projection such as the Azimuthal equidistant projection. There are also projections (Maurer, Close) where true distances from *two* points are preserved.

Choosing a model for the shape of the Earth

Projection construction is also affected by how the shape of the Earth is approximated. In the following discussion on projection categories, a sphere is assumed. However, the Earth is not exactly spherical but is closer in shape to an oblate ellipsoid, a shape which bulges around the equator. Selecting a model for a shape of the Earth involves choosing between the advantages and disadvantages of a sphere versus an ellipsoid.

Spherical models are useful for small-scale maps such as world atlases and globes, since the error at that scale is not usually noticeable or important enough to justify using the more complicated ellipsoid. The ellipsoidal model is commonly used to construct topographic maps and for other large and medium scale maps that need to accurately depict the land surface.

A third model of the shape of the Earth is called a geoid, which is a complex and more or less accurate representation of the global mean sea level surface that is obtained through a combination of terrestrial and satellite gravity measurements. This model is not used for mapping due to its complexity but is instead used for control purposes in the construction of geographic datums. (In geodesy, plural of "datum" is "datums" rather than "data".) A geoid is used to construct a datum by adding irregularities to the ellipsoid in order to better match the Earth's actual shape (it takes into account the large scale features in the Earth's gravity field associated with mantle convection patterns, as well as the gravity signatures of very large geomorphic features such as mountain ranges, plateaus and plains). Historically, datums have been based on ellipsoids that best represent the geoid within the region the datum is intended to map. Each ellipsoid has a distinct major and minor axis. Different controls (modifications) are added to the ellipsoid in order to construct the datum, which is specialized for a specific geographic regions (such as the North American Datum). A few modern datums, such as WGS84 (the one used in the Global Positioning System GPS), are optimized to represent the entire earth as well as possible with a single ellipsoid, at the expense of some accuracy in smaller regions.

Classification

A fundamental projection classification is based on the type of projection surface onto which the globe is conceptually projected. The projections are described in terms of placing a gigantic surface in contact with the earth, followed by an implied scaling operation. These surfaces are cylindrical (e.g. Mercator), conic (e.g., Albers), or azimuthal or plane (e.g. stereographic). Many mathematical projections, however, do not neatly fit into any of these three conceptual projection methods. Hence other peer categories have been described in the literature, such as pseudoconic, pseudocylindrical, pseudoazimuthal, retroazimuthal, and polyconic.

Another way to classify projections is according to properties of the model they preserve. Some of the more common categories are:

- Preserving direction (*azimuthal*), a trait possible only from one or two points to every other point
- Preserving shape locally (*conformal* or *orthomorphic*)
- Preserving area (*equal-area* or *equiareal* or *equivalent* or *authalic*)
- Preserving distance (*equidistant*), a trait possible only between one or two points and every other point
- Preserving shortest route, a trait preserved only by the gnomonic projectio.

NOTE: Because the sphere is not a developable surface, it is impossible to construct a map projection that is both equal-area and conformal.

Projections by Surface

Cylindrical

The term "normal cylindrical projection" is used to refer to any projection in which meridians are mapped to equally spaced vertical lines and circles of latitude (parallels) are mapped to horizontal lines.

The mapping of meridians to vertical lines can be visualized by imagining a cylinder (of which the axis coincides with the Earth's axis of rotation) wrapped around the Earth and then projecting onto the cylinder, and subsequently unfolding the cylinder.

By the geometry of their construction, cylindrical projections stretch distances east-west. The amount of stretch is the same at any chosen latitude on all cylindrical projections, and is given by the secant of the latitude as a multiple of the equator's scale. The various cylindrical projections are distinguished from each other solely by their north-south stretching (where latitude is given by ö):

- North-south stretching is equal to the east-west stretching (secant ö): The east-west scale matches the north-south scale: conformal cylindrical or Mercator; this distorts areas excessively in high latitudes.
- North-south stretching growing rapidly with latitude,

even faster than east-west stretching (secant² ö: The cylindric perspective (= central cylindrical) projection; unsuitable because distortion is even worse than in the Mercator projection.

- North-south stretching grows with latitude, but less quickly than the east-west stretching: such as the Miller cylindrical projection.
- North-south distances neither stretched nor compressed (1): equidistant cylindrical or plate carrée.
- North-south compression precisely the reciprocal of east-west stretching (cosine ö): equal-area cylindrical (with many named specializations such as Gall–Peters or Gall orthographic, Behrmann, and Lambert cylindrical equal-area). This divides north-south distances by a factor equal to the secant of the latitude, preserving area but heavily distorting shapes.

In the first case (Mercator), the east-west scale always equals the north-south scale. In the second case (central cylindrical), the north-south scale exceeds the east-west scale everywhere away from the equator. Each remaining case has a pair of identical latitudes of opposite sign (or else the equator) at which the east-west scale matches the north-south-scale.

Normal cylindrical projections map the whole Earth as a finite rectangle, except in the first two cases, where the rectangle stretches infinitely tall while retaining constant width.

Pseudocylindrical

Pseudocylindrical projections represent the *central* meridian and each parallel as a single straight line segment, but not the other meridians. Each pseudocylindrical projection represents a point on the Earth along the straight line representing its parallel, at a distance which is a function of its difference in longitude from the central meridian.

- Sinusoidal: the north-south scale and the east-west scale are the same throughout the map, creating an equal-area map. On the map, as in reality, the length of each parallel is proportional to the cosine of the latitude. Thus the shape of the map for the whole earth is the region between two symmetric rotated cosine curves.

The true distance between two points on the same meridian corresponds to the distance on the map between the two parallels, which is smaller than the distance between the two points on the map. The true distance between two points on the same parallel – and the true area of shapes on the map – are not distorted. The meridians drawn on the map help the user to realize the shape distortion and mentally compensate for it.

- Collignon projection, which in its most common forms represents each meridian as 2 straight line segments, one from each pole to the equator.
- Mollweide
- Goode homolosine
- Eckert IV
- Eckert VI
- Kavrayskiy VII
- Tobler hyperelliptica.

Hybrid

The HEALPix projection combines an equal-area cylindrical projection in equatorial regions with the Collignon projection in polar areas.

Conical

- Equidistant conic
- Lambert conformal conic
- Albers conic.

Pseudoconical

- Bonne
- Werner cordiform designates a pole and a meridian; distances from the pole are preserved, as are distances from the meridian (which is straight) along the parallels
- Continuous American polyconic.

Azimuthal (Projections onto a Plane)

Azimuthal projections have the property that directions from a central point are preserved (and hence, great circles through the central point are represented by straight lines on the map). Usually these projections also have radial symmetry in the scales and hence in the distortions: map distances from the central point are computed by a function $r(d)$ of the true distance d, independent of the angle; correspondingly, circles with the central point as center are mapped into circles which have as center the central point on the map.

The mapping of radial lines can be visualized by imagining a plane tangent to the Earth, with the central point as tangent point.

The radial scale is $r'(d)$ and the transverse scale $r(d)/(R \sin(d/R))$ where R is the radius of the Earth.

Some azimuthal projections are true perspective projections; that is, they can be constructed mechanically, projecting the surface of the Earth by extending lines from a point of perspective (along an infinite line through the tangent point and the tangent point's antipode) onto the plane:

- The gnomonic projection displays great circles as straight lines. Can be constructed by using a point of perspective at the center of the Earth. $r(d) = c \tan(d/R)$; a hemisphere already requires an infinite map,
- The General Perspective Projection can be constructed by using a point of perspective outside the earth. Photographs of Earth (such as those from the International Space Station) give this perspective.
- The orthographic projection maps each point on the earth to the closest point on the plane. Can be constructed from a point of perspective an infinite distance from the tangent point; $r(d) = c \sin(d/R)$. Can display up to a hemisphere on a finite circle. Photographs of Earth from far enough away, such as the Moon, give this perspective.
- The azimuthal conformal projection, also known as the stereographic projection, can be constructed by using

the tangent point's antipode as the point of perspective. $r(d) = c \tan(d/2R)$; the scale is $c/(2R \cos^2(d/2R))$. Can display nearly the entire sphere on a finite circle. The full sphere requires an infinite map.

Other azimuthal projections are not true perspective projections:

- Azimuthal equidistant: $r(d) = cd$; it is used by amateur radio operators to know the direction to point their antennas toward a point and see the distance to it. Distance from the tangent point on the map is proportional to surface distance on the earth (; for the case where the tangent point is the North Pole, see the flag of the United Nations)
- Lambert azimuthal equal-area. Distance from the tangent point on the map is proportional to straight-line distance through the earth: $r(d) = c \sin(d/2R)$
- Logarithmic azimuthal is constructed so that each point's distance from the center of the map is the logarithm of its distance from the tangent point on the Earth. Works well with cognitive maps. $r(d) = c \ln(d/d_0)$; locations closer than at a distance equal to the constant d_0 are not shown.

Projections by preservation of a metric property

Conformal

Conformal map projections preserve angles locally:

- Mercator-rhumb lines are represented by straight segments
- Stereographic-shape of circles is conserved
- Roussilhe
- Lambert conformal conic
- Quincuncial map
- Adams hemisphere-in-a-square projection
- Guyou hemisphere-in-a-square projection.

Equal-area

These projections preserve area:

- Gall orthographic (also known as Gall–Peters, or Peters, projection)
- Albers conic
- Lambert azimuthal equal-area
- Lambert cylindrical equal-area
- Mollweide
- Hammer
- Briesemeister
- Sinusoidal
- Werner
- Bonne
- Bottomley
- Goode's homolosine
- Hobo-Dyer
- Collignon
- Tobler hyperelliptical.

Equidistant

These preserve distance from some standard point or line:

- Equirectangular-distances along meridians are conserved
- Plate carrée-an Equirectangular projection centred at the equator
- Azimuthal equidistant-distances along great circles radiating from centre are conserved
- Equidistant conic
- Sinusoidal-distances along parallels are conserved
- Werner cordiform distances from the North Pole are correct as are the curved distance on parallels
- Soldner

- Two-point equidistant: two "control points" are arbitrarily chosen by the map maker. Distance from any point on the map to each control point is proportional to surface distance on the earth.

Gnomonic

Great circles are displayed as straight lines:

- Gnomonic projection.

Retroazimuthal

Direction to a fixed location B (the bearing at the starting location A of the shortest route) corresponds to the direction on the map from A to B:

- Littrow-the only conformal retroazimuthal projection
- Hammer retroazimuthal-also preserves distance from the central point
- Craig retroazimuthal *aka* Mecca or Qibla-also has vertical meridians.

Compromise Projections

Compromise projections give up the idea of perfectly preserving metric properties, seeking instead to strike a balance between distortions, or to simply make things "look right". Most of these types of projections distort shape in the polar regions more than at the equator:

- Robinson
- van der Grinten
- Miller cylindrical
- Winkel Tripel
- Buckminster Fuller's Dymaxion
- B.J.S. Cahill's Butterfly Map
- Steve Waterman's Butterfly Map
- Kavrayskiy VII
- Wagner VI
- Chamberlin trimetric
- Oronce Fine's cordiform.

Map Purpose and Informations' Selection

Arthur H. Robinson, an American cartographer influential in thematic cartography, stated that a map not properly designed "will be a cartographic failure." He also claimed, when considering all aspects of cartography, that "map design is perhaps the most complex." Robinson codified the mapmaker's understanding that a map must be designed foremost with consideration to the audience and its needs.

From the very beginning of mapmaking, maps "have been made for some particular purpose or set of purposes". The intent of the map should be illustrated in a manner in which the percipient acknowledges its purpose in a timely fashion. The term *percipient* refers to the person receiving information and was coined by Robinson. The principle of figure-ground refers to this notion of engaging the user by presenting a clear presentation, leaving no confusion concerning the purpose of the map. This will enhance the user's experience and keep his attention. If the user is unable to identify what is being demonstrated in a reasonable fashion, the map may be regarded as useless.

Making a meaningful map is the ultimate goal. Alan MacEachren explains that a well designed map "is convincing because it implies authenticity". An interesting map will no doubt engage a reader. Information richness or a map that is multivariate shows relationships within the map. Showing several variables allows comparison, which adds to the meaningfulness of the map. This also generates hypothesis and stimulates ideas and perhaps further research. In order to convey the message of the map, the creator must design it in a manner which will aid the reader in the overall understanding of its purpose. The title of a map may provide the "needed link" necessary for communicating that message, but the overall design of the map fosters the manner in which the reader interprets it.

In the 21st century it is possible to find a map of virtually anything from the inner workings of the human body to the virtual worlds of cyberspace. Therefore there are now a huge variety of different styles and types of map-for example, one area which has evolved a specific and recognisable variation are those used by public transport organisations to guide passengers, namely urban rail and metro maps, many of which are loosely based on 45 degree angles as originally perfected by Harry Beck and George Dow.

Naming Conventions

Most maps use text to label places and for such things as a map title, legend and other information. Maps are often made in specific languages, though names of places often differ between languages. So a map made in English may use the name *Germany* for that country, while a German map would use *Deutschland* and a French map *Allemagne.* A word that describes a place, using a non-native terminology or language is referred to as an exonym.

In some cases the proper name is not clear. For example, the nation of Burma officially changed its name to Myanmar, but many nations do not recognize the ruling junta and continue to use *Burma.* Sometimes an official name change is resisted in other languages and the older name may remain in common use. Examples include the use of *Saigon* for Ho Chi Minh City, *Bangkok* for Krung Thep and *Ivory Coast* for Côte d'Ivoire.

Difficulties arise, when transliteration or transcription between writing systems is required. National names tend to have well established names in other languages and writing systems, such as *Russia* for Đîññèÿ, but for many placenames a system of transliteration or transcription is required. In transliteration, the symbols of one language are represented by symbols in another.

For example, the Cyrillic letter *Đ* is traditionally written as *R* in the Latin alphabet. Systems exist for transliteration of Arabic, but the results may vary. For example, the Yemeni city of Mocha is written variously in English as Mocha, Al Mukha, al-Mukhâ, Mocca and Moka. Transliteration systems are based on relating written symbols to one another, while transcription is the attempt to spell in one language the phonetic sounds of another. Chinese writing is transformed into the Latin alphabet through the Pinyin phonetic transcription systems. Other systems were used in the past, such as Wade-

Giles, resulting in the city being spelled *Beijing* on newer English maps and *Peking* on older ones.

Further difficulties arise when countries, especially former colonies, do not have a strong national geographic naming standard. In such cases, cartographers may have to choose between various phonetic spellings of local names versus older imposed, sometimes resented, colonial names. Some countries have multiple official languages, resulting in multiple official placenames. For example, the capital of Belgium is both *Brussels* and *Bruxelles*. In Canada, English and French are official languages and places have names in both languages. British Columbia is also officially named *la Colombie-Britannique*. English maps rarely show the French names outside of Quebec, which itself is spelled *Québec* in French.

The study of placenames is called toponymy, while that of the origin and historical usage of placenames as words is etymology.

In order to improve legibility or to aid the illiterate, some maps have been produced using pictograms to represent places. The iconic example of this practice is Lance Wyman's early plans for the Mexico City Metro, on which stations were shown simply as stylized logos. Wyman also prototyped such a map for the Washington Metro, though ultimately the idea was rejected. Other cities experimenting with such maps are Fukuoka, Guadalajara and Monterrey.

Map Symbology

The quality of a map's design affects its reader's ability to extract information and to learn from the map. Cartographic symbology has been developed in an effort to portray the world accurately and effectively convey information to the map reader. A legend explains the pictorial language of the map, known as its symbology. The title indicates the region the map portrays; the map image portrays the region and so on. Although every map element serves some purpose, convention only dictates inclusion of some elements, while others are considered optional. A menu of map elements includes the neatline (border), compass rose or north arrow, overview map, bar scale, projection and information about the map sources, accuracy and publication.

When examining a landscape, scale can be intuited from trees, houses and cars. Not so with a map. Even such a simple thing as a north arrow is crucial. It may seem obvious that the top of a map should point north, but this might not be the case.

Color, likewise, is equally important. How the cartographer displays the data in different hues can greatly affect the understanding or feel of the map. Different intensities of hue portray different objectives the cartographer is attempting to get across to the audience. Today, personal computers can display up to 16 million distinct colors at a time, even though the human eye can distinguish only a minimum number of these (Jeer, 1997). This fact allows for a multitude of color options for even for the most demanding maps. Moreover, computers can easily hatch patterns in colors to give even more options. This is very beneficial, when symbolizing data in categories such as quintile and equal interval classifications.

Quantitative symbols give a visual measure of the relative size/importance/number that a symbol represents and to symbolize this data on a map, there are two major classes of symbols used for portraying quantitative properties. Proportional symbols change their visual weight according to a quantitative property. These are appropriate for extensive statistics. Choropleth maps portray data collection areas, such as counties or census tracts, with color. Using color this way, the darkness and intensity (or value) of the color is evaluated by the eye as a measure of intensity or concentration (Harvard Graduate School of Design, 2005).

Map Generalization

A good map has to provide a compromise between portraying the items of interest (or themes) in the *right place* for the map scale used, against the need to annotate that item with text or a symbol, which takes up space on the map medium and very likely will cause some other item of interest to be displaced. The cartographer is thus constantly making judgements about what to include, what to leave out and what to show in a *slightly* incorrect place-because of the demands of the annotation. This issue assumes more importance as the scale of the map gets smaller (i.e. the map shows a larger area), because relatively, the annotation on the map takes up more space *on the ground*. A good example from the late 1980s was the Ordnance Survey's first digital maps, where the *absolute* positions of major roads shown at scales of 1:1250 and 1:2500 were sometimes a scale distance of hundreds of metres away from ground truth, when shown on digital maps at scales of 1:250000 and 1:625000, because of the overriding need to annotate the features.

Cartographic Errors

Some maps contain deliberate errors or distortions, either as propaganda or as a "watermark" helping the copyright owner identify infringement if the error appears in competitors' maps. The latter often come in the form of nonexistent, misnamed, or misspelled "trap streets". Other names and forms for this are paper townsites, fictitious entries, and copyright easter eggs.

Another motive for deliberate errors is simply cartographic graffiti or prank: a mapmaker wishing to leave his or her mark on the work. Mount Richard, for example, was a fictitious peak on the Rocky Mountains' continental divide that appeared on a Boulder County, Colorado map in the early 1970s. It is believed to be the work of drafts man Richard Ciacci. The fiction was not discovered until two years later

Computational Visualistics

The term Computational visualistics is used for addressing the whole range of investigating scientifically pictures "in" the computer.

Images take a rather prominent place in contemporary life in the western societies. Together with language, they have been connected to human culture from the very beginning. For about one century – after several millennia of written word's dominance – their part is increasing again remarkably. Steps toward a general science of images, which we may call 'general visualistics' in analogy to general linguistics, have only been taken recently. So far, a unique scientific basis for circumscribing and describing the heterogeneous phenomenon "image" in an interpersonally verifiable manner hasstill been missing while distinct aspects falling in the domain of visualistics have predominantly been dealt with in several other disciplines, among them in particular philosophy, psychology, and art history. Last (though not least), important contributions to certain aspects of a new science of images have come from computer science.

In computer science, too, considering pictures evolved originally along several more or less independent questions, which lead to proper sub-disciplines: computer graphics is certainly the most "visible" among them. Only just recently, the effort has been increased to finally form a unique and partially autonomous branch of computer science dedicated to images in general. In analogy to computational linguistics, the artificial expression *computational visualistics* is used for addressing the whole range of investigating scientifically pictures "in" the computer.

Areas Covered

For a science of images within computer science, the abstract data type »image« (or perhaps several such types) stands in the center of interest together with the potential implementations (cf. Schirra 2005). There are three main groups of algorithms for that data type to be considered in computational visualistics:

Algorithms from »Image« to »Image«

In the field called image processing, the focus of attention is formed by the operations that take (at least) one picture (and potentially several secondary parameters that are not images) and relate it to another picture. With these operations, we can define algorithms for improving the quality of images (e.g., contrast reinforcement), and procedures for extracting certain parts of an image (e.g., edge finding) or for stamping out pictorial patterns following a particular Gestalt criterion (e.g., blue screen technique). Compression algorithms for the efficient storing or transmitting of pictorial data also belong into this field.

Algorithms from »Image« to "Not Image"

Two disciplines share the operations transforming images into non-pictorial data items. The field of pattern recognition is actually not restricted to pictures. But it has performed important precursory work for computational visualistics since the early 1950's in those areas that essentially classify information in given images: the identification of simple geometric Gestalts (e.g., "circular region"), the classification of letters (recognition of handwriting), the "seeing" of spatial objects in the images or even the association of stylistic attributes of the representation. That is, the images are to be associated with instances of a non-pictorial data type forming a description of some of their aspects. The neighboring field of computer vision is the part of AI (artificial intelligence) in which computer scientists try to teach – loosely speaking – computers the ability of visual perception. Therefore, a problem rather belongs to computer vision to the degree to which its goal is "semantic", i.e., the result approximates the human seeing of objects in a picture.

Algorithms from "Not-image" to »Image«

The investigation of possibilities gained by the operations that result in instances of the data type »image« but take as a starting point instances of non-pictorial data types is performed in particular in computer graphics and information visualization. The former deals with images in the closer sense, i.e., those pictures showing spatial configurations of objects (in the colloquial meaning of 'object') in a more or less naturalistic representation like, e.g., in virtual architecture. The starting point of the picture-generating algorithms in computer graphics is usually a data type that allows us to describe the geometry in three dimensions and the lighting of the scene to be depicted together with the important optical properties of the surfaces considered.

Scientists in information visualization are interested in presenting pictorially any other data type, in particular those that consist of non-visual components in a "space" of states: in order to do so, a convention of visual presentation has firstly to be determined – e.g., a code of colors or certain icons. The well-known fractal images (e.g., of the Mandelbrot set) form a borderline case of information visualization since an abstract mathematical property has been visualized.

Computational Visualistics Degree Programmes

The subject of computational visualistics was introduced at the University of Magdeburg, Germany, in the fall of 1996. This five-year diploma programme has computer science courses as its core: students learn about digital methods and electronic tools for solving picture-related problems. The technological areas of endeavour are complemented by courses on pictures in the humanities. In addition to learning about the traditional (i.e. not computerized) contexts of using pictures, students intensively practice their communicative skills. As the third component of the program, an application subject such as medicine gives students an early opportunity to apply their knowledge in that they learn the skills needed for co-operating with clients and experts in other fields. Bachelor and master programmes have been introduced in the meantime.

Critical Thinking

Critical thinking, in its broadest sense has been described as "purposeful reflective judgment concerning what to believe or what to do." The list of core critical thinking skills includes interpretation, analysis, inference, evaluation, explanation and meta-cognition. There is a reasonable level of consensus among experts that an individual or group engaged in strong critical thinking gives due consideration to the evidence, the context of judgment, the relevant criteria for making the judgment well, the applicable methods or techniques for forming the judgment, and the applicable theoretical constructs for understanding the problem and the question at hand. In addition to possessing strong critical thinking skills, one must be disposed to engage problems and decisions using those skills. Critical thinking employs not only logic but broad intellectual criteria such as clarity, credibility, accuracy, precision, relevance, depth, breadth, significance and fairness. The positive habits of mind which characterize a person strongly disposed toward critical thinking include a courageous desire to follow reason and evidence wherever they may lead, open-mindedness, foresight attention to the possible consequences of choices, a systematic approach to problem solving, inquisitiveness, fair-mindedness and maturity of judgment, and confidence in reasoning. In reflective problem solving and thoughtful decision making using critical thinking one considers evidence, (like investigating evidence) the context of judgment, the relevant criteria for making the judgment well, the applicable methods or techniques for forming the judgment, and the applicable theoretical constructs for understanding the problem and the question at hand.

"Critical" as used in the expression "critical thinking" connotes the importance or centrality of the thinking to an issue, question or problem of concern. "Critical" in this context does not mean "disapproved" or "negative." There are many positive and useful uses of critical thinking, for example formulating a workable solution to a complex personal problem, deliberating as a group about what course of action to take, or analyzing the assumptions and the quality of the methods used in scientifically arriving at a reasonable level of confidence about a given hypothesis. Using strong critical thinking we might evaluate an argument, for example, as worthy of acceptance because it is valid and based on true premises. Upon reflection, a speaker may be evaluated as a credible source of knowledge on a given topic.

Contemporary cognitive psychology regards human reasoning as a complex process which is both reactive and reflective. The deliberation characteristic of strong critical thinking associates critical thinking with the reflective aspect of human reasoning. Those who would seek to improve our individual and collective capacity to engage problems using strong critical thinking skills are, therefore, recommending that we bring greater reflection and deliberation to decision making. John Dewey is just one of many educational leaders who recognized that a curriculum aimed at building thinking skills would be a benefit not only to the individual learner, but to the community and to the entire democracy. In a seminal study on critical thinking and education in 1941, Edward Glaser writes that the ability to think critically involves three things: 1 An attitude of being disposed (state of mind regarding something) to consider in a thoughtful way the problems and subjects that come within the range of one's experiences, 2 Knowledge of the methods of logical inquiry and reasoning, and

3 Some skill in applying those methods. Educational programs aimed at developing critical thinking in children and adult learners, individually or in group problem solving and decision making contexts, continue to address these same three central elements.

Critical thinking calls for a persistent effort to examine any belief or supposed form of knowledge in the light of the evidence that supports it and the further conclusions to which it tends. It also generally requires ability to recognize problems, to find workable means for meeting those problems, to gather and marshal pertinent (relevant) information, to recognize unstated assumptions and values, to comprehend and use language with accuracy, clarity, and discrimination, to interpret data, to appraise evidence and evaluate arguments, to recognize the existence (or non-existence) of logical relationships between propositions, to draw warranted conclusions and generalizations, to put to test the conclusions and generalizations at which one arrives, to reconstruct one's patterns of beliefs on the basis of wider experience, and to render accurate judgments about specific things and qualities in everyday life.

Critical thinking can occur whenever one judges, decides, or solves a problem; in general, whenever one must figure out what to believe or what to do, and do so in a reasonable and reflective way. Reading, writing, speaking, and listening can all be done critically or uncritically. Critical thinking is crucial to becoming a close reader and a substantive writer. Expressed most generally, critical thinking is "a way of taking up the problems of life." Irrespective of the sphere of thought, "a well cultivated critical thinker":

- raises important questions and problems, formulating them clearly and precisely;
- gathers and assesses relevant information, using abstract ideas to interpret it effectively
- comes to well-reasoned conclusions and solutions, testing them against relevant criteria and standards;
- thinks open-mindfully within alternative systems of thought, recognizing and assessing, as need be, their

assumptions, implications, and practical consequences; and

- communicates effectively with others in figuring out solutions to complex problems; without being unduly influenced by others' thinking on the topic.

Principles and Dispositions

Critical thinking is based on self-corrective concepts and principles, not on hard and fast, or step-by-step, procedures.

Critical thinking employs not only logic (either formal or, much more often, informal) but broad intellectual criteria such as clarity, credibility, accuracy, precision, relevance, depth, breadth, significance.

Critical thinking is an important element of all professional fields and academic disciplines (by referencing their respective sets of permissible questions, evidence sources, criteria, etc.). Within the framework of scientific skepticism, the process of critical thinking involves the careful acquisition and interpretation of information and use of it to reach a well-justified conclusion. The concepts and principles of critical thinking can be applied to any context or case but only by reflecting upon the nature of that application. Critical thinking forms, therefore, a system of related, and overlapping, modes of thought such as anthropological thinking, sociological thinking, historical thinking, political thinking, psychological thinking, philosophical thinking, mathematical thinking, chemical thinking, biological thinking, ecological thinking, legal thinking, ethical thinking, musical thinking, thinking like a painter, sculptor, engineer, business person, etc. In other words, though critical thinking principles are universal, their application to disciplines requires a process of reflective contextualization.

Critical thinking is considered important in the academic fields because it enables one to analyze, evaluate, explain, and restructure their thinking, thereby decreasing the risk of adopting, acting on, or thinking with, a false belief. However, even with knowledge of the methods of logical inquiry and reasoning, mistakes can happen due to a thinker's inability to apply the methods or because of character traits such as egocentrism. Critical thinking includes identification of prejudice, bias, propaganda, self-deception, distortion, misinformation, etc. Given research in cognitive psychology, some educators believe that schools should focus on teaching their students critical thinking skills and cultivation of intellectual traits.

Applications

Critical thinking is about being both willing and able to evaluate one's thinking. Thinking might be criticized because one does not have all the relevant information—indeed, important information may remain undiscovered, or the information may not even be knowable—or because one makes unjustified inferences, uses inappropriate concepts, or fails to notice important implications. One's thinking may be unclear, inaccurate, imprecise, irrelevant, narrow, shallow, illogical, or trivial, due to ignorance or misapplication of the appropriate skills of thinking. On the other hand, one's thinking might be criticized as being the result of a sub-optimal disposition. The dispositional dimension of critical thinking is characterological. Its focus is in developing the habitual intention to be truth-seeking, open-minded, systematic, analytical, inquisitive, confident in reasoning, and prudent in making judgments. Those who are ambivalent on one or more of these aspects of the disposition toward critical thinking, or who have an opposite disposition (intellectually arrogant, bias, intolerant, disorganized, lazy, heedless of consequences, indifferent toward new information, mistrustful of reasoning, or imprudent) are more likely to encounter problems in using their critical thinking skills. Failure to recognize the importance of correct dispositions can lead to various forms of self-deception and closed-mindedness, both individually and collectively.

When individuals possess intellectual skills alone, without the intellectual traits of mind, *weak sense critical thinking* results. Fair-minded or *strong sense critical thinking* requires intellectual humility, empathy, integrity, perseverance, courage, autonomy, confidence in reason, and other intellectual traits. Thus, critical thinking without essential intellectual traits often results in clever, but manipulative and often unethical or subjective thought.

The relationship between critical thinking skills and critical thinking dispositions is an empirical question. Some people have both in abundance, some have skills but not the disposition to use them, some are disposed but lack strong skills, and some have neither. Two measures of critical thinking dispositions are the California Critical Thinking Disposition Inventory and the California Measure of Mental Motivation.

The key to seeing the significance of critical thinking in academics is in understanding the significance of critical thinking in learning. There are two meanings to the learning of this content. The first occurs when learners (for the first time) construct in their minds the basic ideas, principles, and theories that are inherent in content. This is a process of internalization. The second occurs when learners effectively use those ideas, principles, and theories as they become relevant in learners' lives. This is a process of application. Good teachers cultivate critical thinking (intellectually engaged thinking) at every stage of learning, including initial learning. This process of intellectual engagement is at the heart of the Oxford, Durham, Cambridge and London School of Economics tutorials. The tutor questions the students, often in a Socratic manner. The key is that the teacher who fosters critical thinking fosters reflectiveness in students by asking questions that stimulate thinking essential to the construction of knowledge.

As emphasized above, each discipline adapts its use of critical thinking concepts and principles(principles like in school). The core concepts are always there, but they are embedded in subject specific content. For students to learn content, intellectual engagement is crucial. All students must do their own thinking, their own construction of knowledge. Good teachers recognize this and therefore focus on the questions, readings, activities that stimulate the mind to take ownership of key concepts and principles underlying the subject.

In the UK school system, *Critical Thinking* is offered as a subject which 16-to 18-year-olds can take as an A-Level. Under the OCR exam board, students can sit two exam papers for the AS: "Credibility of Evidence" and "Assessing and Developing Argument". The full Advanced GCE is now available: in addition to the two AS units, candidates sit the two papers "Resolution of Dilemmas" and "Critical Reasoning". The A-level tests candidates on their ability to think critically about, and analyze, arguments on their deductive or inductive validity, as well as producing their own arguments. It also tests their ability to analyze certain related topics such as credibility and ethical decision-making. However, due to its comparative lack of subject content, many universities do not accept it as a main A-level for admissions. Nevertheless, the AS is often useful in developing reasoning skills, and the full advanced GCE is useful for degree courses in politics, philosophy, history or theology, providing the skills required for critical analysis that are useful, for example, in biblical study.

There is also an Advanced Extension Award offered in Critical Thinking in the UK, open to any A-level student regardless of whether they have the Critical Thinking A-level. Cambridge International Examinations have an A-level in Thinking Skills. From 2008, Assessment and Qualifications Alliance will also be offering an A-level Critical Thinking specification; OCR exam board have also modified theirs for 2008. Many examinations for university entrance set by universities, on top of A-level examinations, also include a critical thinking component, such as the LNAT, the UKCAT, the BioMedical Admissions Test and the Thinking Skills Assessment.

Research in Efficiency of Critical Thinking Instruction

Research suggests a widespread skepticism about universities' effectiveness in fostering critical thinking. For example, in a three year study of 68 public and private colleges in California, though the overwhelming majority (89%) claimed critical thinking to be a primary objective of their instruction, only a small minority (19%) could give a clear explanation of what critical thinking is. Furthermore, although the overwhelming majority (78%) claimed that their students lacked appropriate intellectual standards (to use in assessing their thinking), and 73% considered that students learning to assess their own work was of primary importance, only a very small minority (8%) could enumerate any intellectual criteria or standards they required of students or could give an intelligible explanation of what those criteria and standards were.

This study mirrors a meta-analysis of the literature on teaching effectiveness in higher education. According to the study, critical reports by authorities on higher education, political leaders and business people have claimed that higher education is failing to respond to the needs of students, and that many of our graduates' knowledge and skills do not meet society's requirements for well-educated citizens. Thus the meta-analysis focused on the question: How valid are these claims? Researchers concluded:

- "Faculty aspire to develop students' thinking skills, but research consistently shows that in practice we tend to aim at facts and concepts in the disciplines, at the lowest cognitive levels, rather than development of intellect or values."
- "Faculty agree almost universally that the development of students' higher-order intellectual or cognitive abilities is the most important educational task of colleges and universities."
- "These abilities underpin our students' perceptions of the world and the consequent decisions they make."
- "Specifically, critical thinking – the capacity to evaluate skillfully and fairly the quality of evidence and detect error, hypocrisy, manipulation, dissembling, and bias – is central to both personal success and national needs."
- A 1972 study of 40,000 faculty members by the American Council on Education found that 97 percent of the respondents indicated the most important goal of undergraduate education is to foster students' ability to think critically.
- Process-oriented instructional orientations "have long been more successful than conventional instruction in fostering effective movement from concrete to formal reasoning. Such programs emphasize students' active

involvement in learning and cooperative work with other students and de-emphasize lectures..."

- "Numerous studies of college classrooms reveal that, rather than actively involving our students in learning, we lecture, even though lectures are not nearly as effective as other means for developing cognitive skills."
- "In addition, students may be attending to lectures only about one-half of their time in class, and retention from lectures is low."
- "Studies suggest our methods often fail to dislodge students' misconceptions and ensure learning of complex, abstract concepts. Capacity for problem solving is limited by our use of inappropriately simple practice exercises."
- "Classroom tests often set the standard for students' learning. As with instruction, however, we tend to emphasize recall of memorized factual information rather than intellectual challenge."
- "Taken together with our preference for lecturing, our tests may be reinforcing our students' commonly fact-oriented memory learning, of limited value to either them or society."

In contrast to these results, for students to excel at thinking critically, especially at the graduate level and in research, where it is crucial, they must be taught not how to know the answer, but how to ask the question. As Schwartz explains in "The importance of stupidity in scientific research," researchers must embrace what they do not know. Critical thinking is a primary tool in approaching this.

Cultivation

There is no simple way to develop the intellectual traits of a critical thinker. One important way requires developing one's intellectual empathy and intellectual humility. The first requires extensive experience in entering and accurately constructing points of view toward which one has negative feelings. The second requires extensive experience in identifying the extent of one's own ignorance in a wide variety of subjects (ignorance whose admission leads one to say, "I thought I *knew*, but I merely *believed*"). One becomes less biased and more broad-minded when one becomes more intellectually empathic and intellectually humble, and that involves time, deliberate practice and commitment. It involves considerable personal and intellectual development.

To develop one's critical thinking traits, one should learn the art of suspending judgment (for example, when reading a novel, watching a movie, engaging in dialogical or dialectical reasoning). Ways of doing this include adopting a perceptive rather than judgmental orientation; that is, avoiding moving from perception to judgment as one applies critical thinking to an issue.

One should become aware of one's own fallibility by:

1. accepting that everyone has subconscious biases, and accordingly questioning any reflexive judgments;
2. adopting an ego-sensitive and, indeed, intellectually humble stance;
3. recalling previous beliefs that one once held strongly but now rejects;
4. tendency towards group think; the amount your belief system is formed by what those around you say instead of what you have personally witnessed;
5. realizing one still has numerous blind spots, despite the foregoing.

An integration of insights from the critical thinking literature and cognitive psychology literature is the "Method of Argument and Heuristic Analysis." This technique illustrates the influences of heuristics and biases on human decision making along with the influences of thinking critically about reasons and claims.

Decision Making

Decision making can be regarded as the mental processes (cognitive process) resulting in the selection of a course of action among several alternatives. Every decision making process produces a final choice. The output can be an action or an opinion of choice. Human performance in decision making terms has been the subject of active research from several perspectives. From a psychological perspective, it is necessary to examine individual decisions in the context of a set of needs, preferences an individual has and values they seek. From a cognitive perspective, the decision making process must be regarded as a continuous process integrated in the interaction with the environment. From a normative perspective, the analysis of individual decisions is concerned with the logic of decision making and rationality and the invariant choice it leads to.

Yet, at another level, it might be regarded as a problem solving activity which is terminated when a satisfactory solution is found. Therefore, decision making is a reasoning or emotional process which can be rational or irrational, can be based on explicit assumptions or tacit assumptions.

Logical decision making is an important part of all science-based professions, where specialists apply their knowledge in a given area to making informed decisions. For example, medical decision making often involves making a diagnosis and selecting an appropriate treatment.

Some research using naturalistic methods shows, however, that in situations with higher time pressure, higher stakes, or increased ambiguities, experts use intuitive decision making rather than structured approaches, following a recognition primed decision approach to fit a set of indicators into the expert's experience and immediately arrive at a satisfactory course of action without weighing alternatives. Recent robust decision efforts have formally integrated uncertainty into the decision making process. However, Decision Analysis, recognized and included uncertainties with a structured and rationally justifiable method of decision making since its conception in 1964.

A major part of decision making involves the analysis of a finite set of alternatives described in terms of some evaluative criteria. These criteria may be benefit or cost in nature. Then the problem might be to rank these alternatives in terms of how attractive they are to the decision maker(s) when all the criteria are considered simultaneously. Another goal might be to just find the best alternative or to determine the relative total priority of each alternative (for instance, if alternatives represent projects competing for funds) when all the criteria are considered simultaneously.

Solving such problems is the focus of multi-criteria decision analysis (MCDA) also known as multi-criteria decision making (MCDM). This area of decision making, although it is very old and has attracted the interest of many researchers and practitioners, is still highly debated as there are many MCDA/ MCDM methods which may yield very different results when they are applied on exactly the same data.

Problem Analysis vs. Decision Making

It's important to differentiate between problem analysis and decision making. The concepts are completely separate from one another. Problem analysis must be done first, then the information gathered in that process may be used towards decision making.

Problem Analysis

- Analyze performance, what should the results be against what they actually are
- Problems are merely deviations from performance standards
- Problem must be precisely identified and described
- Problems are caused by some change from a distinctive feature
- Something can always be used to distinguish between what has and hasn't been effected by a cause
- Causes to problems can be deducted from relevant changes found in analyzing the problem
- Most likely cause to a problem is the one that exactly explains all the facts.

Decision Making

- Objectives must first be established

- Objectives must be classified and placed in order of importance
- Alternative actions must be developed
- The alternative must be evaluated against all the objectives
- The alternative that is able to achieve all the objectives is the tentative decision
- The tentative decision is evaluated for more possible consequences
- The decisive actions are taken, and additional actions are taken to prevent any adverse consequences from becoming problems and starting both systems (problem analysis and decision making) all over again.

Everyday Techniques

Some of the decision making techniques people use in everyday life include:

- Listing the advantages and disadvantages of each option, popularized by Plato and Benjamin Franklin
- Choosing the alternative with the highest probability-weighted utility for each alternative
- Satisficing: Accepting the first option that seems like it might achieve the desired result
- Acquiesce to a person in authority or an "expert", just following orders
- Flipism: Flipping a coin, cutting a deck of playing cards, and other random or coincidence methods
- Prayer, tarot cards, astrology, augurs, revelation, or other forms of divination.

Cognitive and Personal Biases

Biases can creep into our decision making processes. Many different people have made a decision about the same question (*e.g.* "Should I have a doctor look at this troubling breast cancer symptom I've discovered?" "Why did I ignore the evidence that the project was going over budget?") and then craft potential cognitive interventions aimed at improving decision making outcomes. Below is a list of some of the more commonly debated cognitive biases.

- Selective search for evidence (a.k.a. Confirmation bias in psychology) (Scott Plous, 1993) – We tend to be willing to gather facts that support certain conclusions but disregard other facts that support different conclusions. Individuals who are highly defensive in this manner show significantly greater left prefrontal cortex activity as measured by EEG than do less defensive individuals.
- Premature termination of search for evidence – We tend to accept the first alternative that looks like it might work.
- Inertia – Unwillingness to change thought patterns that we have used in the past in the face of new circumstances.
- Selective perception – We actively screen-out information that we do not think is important. In one demonstration of this effect, discounting of arguments with which one disagrees (by judging them as untrue or irrelevant) was decreased by selective activation of right prefrontal cortex.
- Wishful thinking or optimism bias – We tend to want to see things in a positive light and this can distort our perception and thinking.
- Choice-supportive bias occurs when we distort our memories of chosen and rejected options to make the chosen options seem more attractive.
- Recency – We tend to place more attention on more recent information and either ignore or forget more distant information. The opposite effect in the first set of data or other information is termed Primacy effect (Plous, 1993).
- Repetition bias – A willingness to believe what we have been told most often and by the greatest number of different sources.

- Anchoring and adjustment – Decisions are unduly influenced by initial information that shapes our view of subsequent information.
- Group think – Peer pressure to conform to the opinions held by the group.
- Source credibility bias – We reject something if we have a bias against the person, organization, or group to which the person belongs: We are inclined to accept a statement by someone we like.
- Incremental decision making and escalating commitment – We look at a decision as a small step in a process and this tends to perpetuate a series of similar decisions. This can be contrasted with zero-based decision making.
- Attribution asymmetry – We tend to attribute our success to our abilities and talents, but we attribute our failures to bad luck and external factors. We attribute other's success to good luck, and their failures to their mistakes.
- Role fulfillment (Self Fulfilling Prophecy) – We conform to the decision making expectations that others have of someone in our position.
- Underestimating uncertainty and the illusion of control – We tend to underestimate future uncertainty because we tend to believe we have more control over events than we really do. We believe we have control to minimize potential problems in our decisions.

Post Decision Analysis

Evaluation and analysis of past decisions is complementary to decision making;

Cognitive Styles

Influence of Briggs Myers Type

According to behavioralist Isabel Briggs Myers, a person's decision making process depends to a significant degree on their cognitive style. Myers developed a set of four bi-polar dimensions, called the Myers-Briggs Type Indicator (MBTI). The terminal points on these dimensions are: *thinking* and *feeling*; *extroversion* and *introversion*; *judgment* and *perception*; and *sensing* and *intuition*. She claimed that a person's decision making style correlates well with how they score on these four dimensions. For example, someone who scored near the thinking, extroversion, sensing, and judgment ends of the dimensions would tend to have a logical, analytical, objective, critical, and empirical decision making style. However, some psychologists say that the MBTI lacks reliability and validity and is poorly constructed.

Other studies suggest that these national or cross-cultural differences exist across entire societies. For example, Maris Martinsons has found that American, Japanese and Chinese business leaders each exhibit a distinctive national style of decision making.

Optimizing vs. Satisficing

Herbert Simon coined the phrase "bounded rationality" to express the idea that human decision-making is limited by available information, available time, and the information-processing ability of the mind. Simon also defined two cognitive styles: *maximizers* try to make an optimal decision, whereas *satisficers* simply try to find a solution that is "good enough". Maximizers tend to take longer making decisions due to the need to maximize performance across all variables and make tradeoffs carefully; they also tend to more often regret their decisions.

Combinatoral vs. Positional

Styles and methods of decision making were elaborated by the founder of Predispositioning Theory, Aron Katsenelinboigen. In his analysis on styles and methods Katsenelinboigen referred to the game of chess, saying that "chess does disclose various methods of operation, notably the creation of predisposition—methods which may be applicable to other, more complex systems."

In his book Katsenelinboigen states that apart from the methods (reactive and selective) and sub-methods (randomization, predispositioning, programming), there are two major styles – positional and combinational. Both styles are utilized in the game of chess. According to Katsenelinboigen, the two styles reflect two basic approaches to the uncertainty: deterministic (combinational style) and indeterministic (positional style). Katsenelinboigen's definition of the two styles are the following.

The combinational style is characterized by;

- a very narrow, clearly defined, primarily material goal, and
- a program that links the initial position with the final outcome.

In defining the combinational style in chess, Katsenelinboigen writes:

The combinational style features a clearly formulated limited objective, namely the capture of material (the main constituent element of a chess position). The objective is implemented via a well defined and in some cases in a unique sequence of moves aimed at reaching the set goal. As a rule, this sequence leaves no options for the opponent. Finding a combinational objective allows the player to focus all his energies on efficient execution, that is, the player's analysis may be limited to the pieces directly partaking in the combination. This approach is the crux of the combination and the combinational style of play.

The positional style is distinguished by;

- a positional goal and
- a formation of semi-complete linkages between the initial step and final outcome.

"Unlike the combinational player, the positional player is occupied, first and foremost, with the elaboration of the position that will allow him to develop in the unknown future. In playing the positional style, the player must evaluate relational and material parameters as independent variables. (...) The positional style gives the player the opportunity to develop a position until it becomes pregnant with a combination. However, the combination is not the final goal of the positional player—

it helps him to achieve the desirable, keeping in mind a predisposition for the future development. The Pyrrhic victory is the best example of one's inability to think positionally."

The positional style serves to;

a) create a predisposition to the future development of the position;
b) induce the environment in a certain way;
c) absorb an unexpected outcome in one's favor;
d) avoid the negative aspects of unexpected outcomes.

The positional style gives the player the opportunity to develop a position until it becomes pregnant with a combination. Katsenelinboigen writes:

> *"As the game progressed and defence became more sophisticated the combinational style of play declined.... The positional style of chess does not eliminate the combinational one with its attempt to see the entire program of action in advance. The positional style merely prepares the transformation to a combination when the latter becomes feasible."*

Neuroscience Perspective

The anterior cingulate cortex (ACC), orbitofrontal cortex (and the overlapping ventromedial prefrontal cortex) are brain regions involved in decision making processes. A recent neuroimaging study, found distinctive patterns of neural activation in these regions depending on whether decisions were made on the basis of personal volition or following directions from someone else. Patients with damage to the ventromedial prefrontal cortex have difficulty making advantageous decisions.

A recent study involving Rhesus monkeys found that neurons in the parietal cortex not only represent the formation of a decision but also signal the degree of certainty (or "confidence") associated with the decision. Another recent study found that lesions to the ACC in the macaque resulted in impaired decision making in the long run of reinforcement guided tasks suggesting that the ACC may be involved in evaluating past reinforcement information and guiding future action.

Emotion appears to aid the decision making process: Decision making often occurs in the face of uncertainty about whether one's choices will lead to benefit or harm. The somatic-marker hypothesis is a neurobiological theory of how decisions are made in the face of uncertain outcome. This theory holds that such decisions are aided by emotions, in the form of bodily states, that are elicited during the deliberation of future consequences and that mark different options for behaviour as being advantageous or disadvantageous. This process involves an interplay between neural systems that elicit emotional/ bodily states and neural systems that map these emotional/ bodily states.

Although it is unclear whether the studies generalize to all processing, there is evidence that volitional movements are initiated, not by the conscious decision making self, but by the subconscious.

Diagrammatic Reasoning

Diagrammatic reasoning is reasoning by means of visual representations. The study of *diagrammatic reasoning* is about the understanding of concepts and ideas, visualized with the use of diagrams and imagery instead of by linguistic or algebraic means.

Characteristica Universalis

Characteristica universalis, commonly interpreted as *universal characteristic*, or *universal character* in English, is a universal and formal language imagined by the German philosopher Gottfried Leibniz able to express mathematical, scientific, and metaphysical concepts. Leibniz thus hoped to create a language usable within the framework of a universal logical calculation or *calculus ratiocinator*.

Since the *characteristica universalis* is diagrammatic and employs pictograms (below left), the diagrams in Leibniz's work warrant close study. On at least two occasions, Leibniz illustrated his philosophical reasoning with diagrams. One diagram, the frontispiece to his 1666 *De Arte Combinatoria* (On the Art of Combinations), represents the Aristotelian theory of how all material things are formed from combinations of the elements earth, water, air, and fire.

These four elements make up the four corners of a diamond. Opposing pairs of these are joined by a bar labeled "contraries" (earth-air, fire-water). At the four corners of the superimposed square are the four qualities defining the elements. Each adjacent pair of these is joined by a bar labeled "possible combination"; the diagonals joining them are labeled "impossible combination." Starting from the top, fire is formed from the combination of dryness and heat; air from wetness and heat; water from coldness and wetness; earth from coldness and dryness.

Diagram

A diagram is a 2D geometric symbolic representation of information according to some visualization technique. Sometimes, the technique uses a 3D visualization which is then projected onto the 2D surface. The term diagram in common sense can have two meanings.

- *visual information device*: Like the term "illustration" the diagram is used as a collective term standing for the whole class of technical genres, including graphs, technical drawings and tables.
- *specific kind of visual display*: This is only the genre, that shows qualitative data with shapes that are connected by lines, arrows, or other visual links.

In science you will find the term used in both ways. For example Anderson (1997) stated more general "diagrams are pictorial, yet abstract, representations of information, and maps, line graphs, bar charts, engineering blueprints, and architects' sketches are all examples of diagrams, whereas photographs and video are not". On the other hand Lowe (1993) defined diagrams as specifically "abstract graphic portrayals of the subject matter they represent".

In the specific sense diagrams and charts contrast computer graphics, technical illustrations, infographics, maps, and technical drawings, by showing "abstract rather than literal representations of information". The essences of a diagram can be seen as:

- a *form* of visual formatting devices
- a *display* that does not show quantitative data, but rather relationships and abstract information
- with *building blocks* such as geometrical shapes that are connected by lines, arrows, or other visual links.

Or in Hall's (1996) words "diagrams are simplified figures, caricatures in a way, intended to convey essential meaning". According to Jan V. White (1984) "the characteristics of a good diagram are elegance, clarity, ease, pattern, simplicity, and validity". Elegance for White means that what you are seeing in the diagram is "the simplest and most fitting solution to a problem".

Logical Graph

A logical graph is a special type of graph-theoretic structure in any one of several systems of graphical syntax that Charles Sanders Peirce developed for logic.

In his papers on *qualitative logic*, *entitative graphs*, and *existential graphs*, Peirce developed several versions of a graphical formalism, or a graph-theoretic formal language, designed to be interpreted for logic.

In the century since Peirce initiated this line of development, a variety of formal systems have branched out from what is abstractly the same formal base of graph-theoretic structures.

Conceptual Graph

A conceptual graph (CG) is a notation for logic based on the existential graphs of Charles Sanders Peirce and the semantic networks of artificial intelligence. In the first published paper on conceptual graphs, John F. Sowa used them to represent the conceptual schemas used in database systems.

His first book applied them to a wide range of topics in artificial intelligence, computer science, and cognitive science. A linear notation, called the *Conceptual Graph Interchange*

Format (CGIF), has been standardized in the ISO standard for Common Logic. The diagram on the right is an example of the *display form* for a conceptual graph. Each box is called a *concept node*, and each oval is called a *relation node*. In CGIF, this CG would be represented by the following statement:

[Cat Elsie] [Sitting *x] [Mat *y] (agent ?x Elsie) (location ?x ?y)

In CGIF, brackets enclose the information inside the concept nodes, and parentheses enclose the information inside the relation nodes. The letters x and y, which are called *coreference labels*, show how the concept and relation nodes are connected. In the *Common Logic Interchange Format (CLIF)*, those letters are mapped to variables, as in the following statement:

(exists ((x Sitting) (y Mat)) (and (Cat Elsie) (agent x Elsie) (location x y)))

As this example shows, the asterisks on the coreference labels *x and *y in CGIF map to existentially quantified variables in CLIF, and the question marks on ?x and ?y map to bound variables in CLIF. A universal quantifier, represented *@every*z* in CGIF, would be represented *forall (z)* in CLIF.

Entitativc Graph

An entitative graph is an element of the graphical syntax for logic that Charles Sanders Peirce developed under the name of *qualitative logic* beginning in the 1880s, taking thc coverage of the formalism only as far as the propositional or sentential aspects of logic are concerned.

The syntax is:

- The blank page;
- Single letters, phrases;
- Objects (subgraphs) enclosed by a simple closed curve called a *cut*. A cut can be empty.

The semantics are:

- The blank page denotes *False*;
- Letters, phrases, subgraphs, and entire graphs can be *True'* or *False*;

- To surround objects with a cut is equivalent to Boolean complementation. Hence an empty cut denotes *Truth*;
- All objects within a given cut are tacitly joined by disjunction.

A "proof" manipulates a graph, using a short list of rules, until the graph is reduced to an empty cut or the blank page. A graph that can be so reduced is what is now called a tautology (or the complement thereof). Graphs that cannot be simplified beyond a certain point are analogues of the satisfiable formulas of first-order logic.

Existential Graph

An existential graph is a type of diagrammatic or visual notation for logical expressions, proposed by Charles Sanders Peirce, who wrote his first paper on graphical logic in 1882 and continued to develop the method until his death in 1914. Peirce proposed three systems of existential graphs:

- *alpha* – isomorphic to sentential logic and the two-element Boolean algebra;
- *beta* – isomorphic to first-order logic with identity, with all formulas closed;
- *gamma* – (nearly) isomorphic to normal modal logic.

Alpha nests in *beta* and *gamma*. *Beta* does not nest in *gamma*, quantified modal logic being more than even Peirce could envisage.

In *alpha* the syntax is:

- The blank page;
- Single letters or phrases written anywhere on the page;
- Any graph may be enclosed by a simple closed curve called a *cut* or *sep*. A cut can be empty. Cuts can nest and concatenate at will, but must never intersect.

Any well-formed part of a graph is a *subgraph*.

The semantics are:

- The blank page denotes *Truth*;
- Letters, phrases, subgraphs, and entire graphs may be *True* or *False*;

- To enclose a subgraph with a cut is equivalent to logical negation or Boolean complementation. Hence an empty cut denotes *False*;
- All subgraphs within a given cut are tacitly conjoined.

Hence the *alpha* graphs are a minimalist notation for sentential logic, grounded in the expressive adequacy of *And* and *Not*. The *alpha* graphs constitute a radical simplification of the two-element Boolean algebra and the truth functors.

The Venn-II Reasoning System

In the early 1990s Sun-Joo Shin presented an extension of Existential Graphs called Venn-II. Syntax and semantics are given formally, together with a set of *Rules of Transformation* which are shown to be sound and complete. Proofs proceed by applying the rules (which remove or add syntactic elements to or from diagrams) sequentially. Venn-II is equivalent in expressive power to a first-order monadic language.

Geovisualization

Geovisualization, short for *Geographic Visualization*, refers to a set of tools and techniques supporting geospatial data analysis through the use of interactive visualization.

Like the related fields of scientific visualization and information visualization geovisualization emphasizes knowledge construction over knowledge storage or information transmission. To do this, geovisualization communicates geospatial information in ways that, when combined with human understanding, allow for data exploration and decision-making processes.

Traditional, static maps have a limited exploratory capability; the graphical representations are inextricably linked to the geographical information beneath. GIS and geovisualization allow for more interactive maps; including the ability to explore different layers of the map, to zoom in or out, and to change the visual appearance of the map, usually on a computer display. Geovisualization represents a set of cartographic technologies and practices that take advantage of the ability of modern microprocessors to render changes to a map in real time, allowing users to adjust the mapped data on the fly.

History

The term visualization is first mentioned in the cartographic literature at least as early as 1953, in an article by University of Chicago geographer Allen K. Philbrick. New developments in the field of computer science prompted the National Science Foundation to redefine the term in a 1987 report which placed visualization at the convergence of computer graphics, image processing, computer vision, computer-aided design, signal processing, and user interface studies and emphasized both the knowledge creation and hypothesis generation aspects of scientific visualization.

Geovisualization developed as a field of research in the early 1980s, based largely on the work of French graphic theorist Jacques Bertin. Bertin's work on cartographic design and information visualization share with the National Science Foundation report a focus on the potential for the use of "dynamic visual displays as prompts for scientific insight and on the methods through which dynamic visual displays might leverage perceptual cognitive processes to facilitate scientific thinking".

Geovisualization has continued to grow as a subject of practice and research. The International Cartographic Association (ICA) established a Commission on Visualization & Virtual Environments in 1995.

Related Fields

Geovisualization is closely related to other visualization fields, such as scientific visualization and information visualization. Owing to its roots in cartography, geovisualization contributes to these other fields by way of the map metaphor, which "has been widely used to visualize non-geographic information in the domains of information visualization and domain knowledge visualization. It is also related to urban simulation.

Practical Applications

Geovisualization has made inroads in a diverse set of real-world situations calling for the decision-making and knowledge creation processes it can provide. The following list provides a summary of some of these applications as they are discussed in the geovisualization literature.

Forestry

Geovisualizers, working with European foresters, used CommonGIS and Visualization Toolkit (VTK) to visualize a large set of spatio-temporal data related to European forests, allowing the data to be explored by non-experts over the Internet. The report summarizing this effort "uncovers a range of fundamental issues relevant to the broad field of geovisualization and information visualization research".

The research team cited the two major problems as the inability of the geovisualizers to convince the foresters of the efficacy of geovisualization in their work and the foresters' misgivings over the dataset's accessibility to non-experts engaging in "uncontrolled exploration". While the geovisualizers focused on the ability of geovisualization to aid in knowledge construction, the foresters preferred the information-communication role of more traditional forms of cartographic representation.

Archaeology

Geovisualization provides archaeologists with a potential technique for mapping unearthed archaeological environments as well as for accessing and exploring archaeological data in three dimensions.

The implications of geovisualization for archaeology are not limited to advances in archaeological theory and exploration but also include the development of new, collaborative relationships between archaeologists and computer scientists.

Environmental Studies

Geovisualization tools provide multiple stakeholders with the ability to make balanced environmental decisions by taking into account the "the complex interacting factors that should be taken into account when studying environmental changes". Geovisualization users can use a georeferenced model to explore a complex set of environmental data, interrogating a number of scenarios or policy options to determine a best fit.

Urban Planning

Both planners and the general public can use geovisualization to explore real-world environments and model 'what if' scenarios based on spatio-temporal data.

While geovisualization in the preceding fields may be divided into two separate domains—the private domain, in which professionals use geovisualization to explore data and generate hypotheses, and the public domain, in which these professionals present their "visual thinking" to the general public —planning relies more heavily than many other fields on collaboration between the general public and professionals.

Planners use geovisualization as a tool for modelling the environmental interests and policy concerns of the general public. Jiang et al. mention two examples, in which "3D photorealistic representations are used to show urban redevelopment [and] dynamic computer simulations are used to show possible pollution diffusion over the next few years."

The widespread use of the Internet by the general public has implications for these collaborative planning efforts, leading to increased participation by the public while decreasing the amount of time it takes to debate more controversial planning decisions.

Traffic Analysis

Traffic analysis is the process of intercepting and examining messages in order to deduce information from patterns in communication. It can be performed even when the messages are encrypted and cannot be decrypted. In general, the greater the number of messages observed, or even intercepted and stored, the more can be inferred from the traffic. Traffic analysis can be performed in the context of military intelligence or counter-intelligence, and is a concern in computer security.

Traffic analysis tasks may be supported by dedicated computer software programs, including commercially available programs such as those offered by i2, Visual Analytics, Memex,

Orion Scientific, Pacific Northwest National Labs, Genesis EW's GenCOM Suite and others. Advanced traffic analysis techniques may include various forms of social network analysis.

In Military Intelligence

In a military context, traffic analysis is a basic part of signals intelligence, and can be a source of information about the intentions and actions of the enemy. Representative patterns include:

- Frequent communications — can denote planning
- Rapid, short, communications — can denote negotiations
- A lack of communication — can indicate a lack of activity, or completion of a finalized plan
- Frequent communication to specific stations from a central station — can highlight the chain of command
- Who talks to whom — can indicate which stations are 'in charge' or the 'control station' of a particular network. This further implies something about the personnel associated with each station
- Who talks when — can indicate which stations are active in connection with events, which implies something about the information being passed and perhaps something about the personnel/access of those associated with some stations
- Who changes from station to station, or medium to medium — can indicate movement, fear of interception

There is a close relationship between traffic analysis and cryptanalysis (commonly called codebreaking). Callsigns and addresses are frequently encrypted, requiring assistance in identifying them.

Traffic volume can often be a sign of an addressee's importance, giving hints to pending objectives or movements to cryptanalysts.

Traffic Flow Security

Traffic-flow security is the use of measures that conceal the presence and properties of valid messages on a network to prevent traffic analysis. This can be done by operational procedures or by the protection resulting from features inherent in some cryptographic equipment. Techniques used include:

- changing radio callsigns frequently
- encryption of a message's sending and receiving addresses (codress messages)
- causing the circuit to appear busy at all times or much of the time by sending dummy traffic
- sending a continuous encrypted signal, whether or not traffic is being transmitted. This is also called masking or link encryption.

Traffic-flow security is one aspect of communications security.

COMINT Metadata Analysis

The Communications' Metadata Intelligence, or COMINT metadata is a term in COMINT referring to the concept of producing intelligence by analyzing only the technical metadata, hence, is a great practical example for traffic analysis in intelligence.

While traditionally information gathering in COMINT is derived from intercepting transmissions, tapping the target's communications and monitoring the content of conversations, the metadata intelligence is not based on content but on technical communicational data.

Non content COMINT is usually used to figure information about the user of a certain transmitter, such as locations, contacts, activity volume, routine and its exceptions.

Examples

For example, if a certain emitter is known as the radio transmitter of a certain unit, and by using DF (direction finding) tools, the position of the emitter is locatable; hence the changes of locations can be monitored

That way we're able to understand that this certain unit is moving from one point to another, without listening to any orders or reports.

If we know that this unit reports back to a command on a certain pattern, and we know that another unit reports on the same pattern to the same command, then the two units are probably related, and that conclusion is based on the metadata of the two units' transmissions, and not on the content of their transmissions.

Using all, or as much of the metadata available is commonly used to build up an Electronic Order of Battle (EOB) – mapping different entities in the battlefield and their connections. Of course the EOB could be built by tapping all the conversations and trying to understand which unit is where, but using the metadata with an automatic analysis tool enables a much faster and accurate EOB build-up that alongside tapping builds a much better and complete picture.

World War I

- British analysts in World War I noticed that the call sign of German Vice Admiral Reinhard Scheer, commanding the hostile fleet, had been transferred to a land-based station. Admiral Beattie, ignorant of Scheer's practice of changing callsigns upon leaving harbor, dismissed its importance and disregarded Room 40 analysts' attempts to make the point. The German fleet sortied, and the British were late in meeting them at the Battle of Jutland. If traffic analysis had been taken more seriously, the British might have done better than a 'draw'.

World War II

- In early World War II, the aircraft carrier HMS *Glorious* was evacuating pilots and planes from Norway. Traffic analysis produced indications *Scharnhorst* and *Gneisenau* were moving into the North Sea, but the Admiralty dismissed the report as unproven. The captain of *Glorious* did not keep sufficient lookout, and was subsequently surprised and sunk. Harry Hinsley, the young Bletchley Park liaison to the Admiralty, later said his reports from the traffic analysts were taken much more seriously thereafter.

- During the planning and rehearsal for the attack on Pearl Harbor, very little traffic passed by radio, subject to interception. The ships, units, and commands involved were all in Japan and in touch by phone, courier, signal lamp, or even flag. None of that traffic was intercepted, and could not be analyzed.
- The espionage effort against Pearl Harbor before December didn't send an unusual number of messages; Japanese vessels regularly called in Hawaii and messages were carried aboard by consular personnel. At least one such vessel carried some Japanese Navy Intelligence officers. Such messages cannot be analyzed. It has been suggested, however, the volume of diplomatic traffic to and from certain consular stations might have indicated places of interest to Japan, which might thus have suggested locations to concentrate traffic analysis and decryption efforts.
- Admiral Nagumo's Pearl Harbor Attack Force sailed under radio silence, with its radios physically locked down. It is unclear if this deceived the U.S.; Pacific Fleet intelligence was unable to locate the Japanese carriers in the days immediately preceding the attack on Pearl Harbor (Kahn).
- The Japanese Navy played radio games to inhibit traffic analysis with the attack force after it sailed in late November. Radio operators normally assigned to carriers, with a characteristic Morse Code "fist", transmitted from inland Japanese waters, suggesting the carriers were still near Japan (Kahn)
- Operation Quicksilver, part of the British deception plan for the Invasion of Normandy in World War II, fed German intelligence a combination of true and false information about troop deployments in Britain, causing the Germans to deduce an order of battle which suggested an invasion at the Pas-de-Calais instead of Normandy. The fictitious divisions created for this deception were supplied with real radio units, which maintained a flow of messages consistent with the deception.

In Computer Security

Traffic analysis is also a concern in computer security. An attacker can gain important information by monitoring the frequency and timing of network packets. A timing attack on the SSH protocol can use timing information to deduce information about passwords since, during interactive session, SSH transmits each keystroke as a message. The time between keystroke messages can be studied using hidden Markov models. Song, *et al.* claim that it can recover the password fifty times faster than a brute force attack.

Onion routing systems are used to gain anonymity. Traffic analysis can be used to attack anonymous communication systems like the Tor anonymity network. Steven J. Murdoch and George Danezis from University of Cambridge presented research showing that traffic-analysis allows adversaries to infer which nodes relay the anonymous streams. This reduces the anonymity provided by Tor. They have shown that otherwise unrelated streams can be linked back to the same initiator.

Remailer systems can also be attacked via traffic analysis. If a message is observed going to a remailing server, and an identical-length (if now anonymized) message is seen exiting the server soon after, a traffic analyst may be able (automatically) connect the sender with the ultimate receiver. Variations of remailer operations exist that can make traffic analysis less effective.

Countermeasures

It is difficult to defeat traffic analysis without both encrypting messages and masking the channel. When no actual messages are being sent, the channel can be masked by sending dummy traffic, similar to the encrypted traffic, thereby keeping bandwidth usage constant. "It is very hard to hide information about the size or timing of messages. The known solutions require Alice to send a continuous stream of messages at the maximum bandwidth she will ever use...This might be acceptable for military applications, but it is not for most civilian applications." The military-versus-civilian problems applies in situations where the user is charged for the volume of information sent.

Even for Internet access, where there is not a per-packet charge, ISPs make statistical assumption that connections from user sites will not be busy 100% of the time. The user cannot simply increase the bandwidth of the link, since masking would fill that as well. If masking, which often can be built into end-to-end encryptors, becomes common practice, ISPs will have to change their traffic assumptions.

7

Point Pattern Analysis

Point Pattern Analysis involves the ability to describe patterns of locations of point events and test whether there is a significant occurrence of clustering of points in a particular area.

Overview

Historically, Point Pattern Analysis was first noted in the works of botanists and ecologists in the 1930s (Chakravorty, 1995). However, in the intervening years, many different fields have also started to use point pattern analysis, such as archeology, epidemiology, astronomy, and criminology. In general, Point Pattern Analysis can be used to describe any type of incident data.

For instance, we may want to conduct "Hot Spot" analysis in order to better understand locations of crimes, or else we may want to study breakouts of certain diseases to better see whether there is a pattern. In both of these cases, Point Pattern Analysis can be of great help to institutions and policymakers in their decisions on how to best allocate their scarce resources to different areas.

Criteria

In order to conduct Point Pattern Analysis, your data must meet five important criteria:

1. The pattern must be mapped on a plane, meaning that you will need both latitude and longitude coordinates.

2. A study area must be selected and determined prior to the analysis.
3. The Point Data should not be a selected sample, but rather the entire set of data you seek to analyze.
4. There should be a one-to-one correspondence between objects in the study area and events in the pattern.
5. The Points must be true incidents with real spatial coordinates. For example, using the centroids of a census tract would not be an especially useful process.

Techniques to Analyze Point Pattern Data

When we are examining incident data, we often need to first get the coordinates of each incident and determine the study area that we wish to use. For instance, if we were examining one hundred robberies within a square mile, we would not want to use a study area of 5 square miles. Although this may sound obvious, we also want to examine our data and make sure that we are not estimating beyond areas, for which we have no data. In general, when we are examining areas to see whether incidents are clustered we are using a null hypothesis that there is no clustering present and that incidents are evenly spread throughout the study area. Sometimes, we may specify that incidents are evenly clustered, controlling for certain variables, such as population density. In general there are Three Types of Techniques: 1.) Quadrant Count Methods, 2.) Kernel Density Estimation (sometimes called K-Means), and 3.) Nearest Neighbor Distance 1.) Quadrant Count Methods: This method involves simply recording and counting the number of events that occur in each quadrant.

In general, it is important to remember that large quadrants produce a very coarse description of the pattern, but as quadrant size is reduced, many areas may become too small and some may contain no events at all.

Two examples of this type of point pattern analysis are Mode and Fuzzy Mode. 2.) Kernel Density Estimation: This method counts the incidents in an area (a kernel), centred at the location where the estimate is made. This analysis is a partitioning technique, meaning that incidents are partitioned into a number of different clusters. Oftentimes the user is able to specify the number of clusters. In some forms of this analysis, all incidents, even the outliers, are assigned to one and only one group. However, other techniques allow for a form of "clumping" analysis, where there are groups that have overlapping membership.

This method is very good for analyzing the point patterns to discover the Hot Spots.

Also, this method provides us with a useful link to geographical data because it is able to transform our data into a density surface.

Our choice of r, the kernel bandwidth strongly affects our density surface.

Also, we can weight these patterns with other data – such as density of populations and unemployment rates.

In Dual Kernel Estimates, you are able to weight the estimates against another set of incidents. For instance you might want to analyze the number of assaults against establishments that are allowed to serve liquor. 3.) Nearest-Neighbor Distance: This method measures the distance from one point to the nearest neighbor point. In general there are three different functions that users are able to employ in Nearest Neighbor Analyses:

G Function: This is the simplest measure and is similar to the mean, however instead of summarizing with a mean, the G function allows us to examine the cumulative frequency distribution of the nearest neighbor distances. The shape of this function can tell us a lot about the way the events are clustered in a point pattern. If events are clustered together, G increases rapidly at short distances, and if events are evenly spaced, G increases slowly up to the distance at which most events are spaced, and only then increases rapidly.

F Function: Instead of accumulating the fraction of the nearest-neighbor distances between events, this measure selects point locations anywhere in the study region at random, and the minimum distance from them to any event in the pattern is determined. 3. K Function: Imagine placing circles of a defined radius centred on the event in turn. Then, the number of events inside the circle's radius is totaled, and the mean count for all of the incidents is totaled. This mean count is then divided by the overall study area. Because all of the incidents are used, the K function provides more information about patterns and clusters then either G or F.

Spatial Point Pattern Analysis and its Application in Geographical Epidemiology

The analysis of spatial point patterns came to prominence in geography during the late 1950s and early 1960s, when a spatial analysis paradigm began to take firm hold within the discipline. Researchers borrowed freely from the plant ecology literature, adopting techniques that had been used there in the description of spatial patterns and applying them in other contexts: for example, in studies of settlement distributions (Dacey 1962; King 1962), the spatial arrangement of stores within urban areas (Rogers 1965) and the distribution of drumlins in glaciated areas (Trenhaile 1971). The methods that were used could be classified into two broad types (Haggett *et al.* 1977).

The first were *distance-based* techniques, using information on the spacing of the points to characterize pattern (typically, mean distance to the nearest neighbouring point). Other techniques were *area-based*, relying on various characteristics of the frequency distribution of the observed numbers of points in regularly defined sub-regions of the study area ('quadrats').

For many geographers, point pattern analysis will conjure up images of 'nearest-neighbour analysis' applied inappropriately to data sets of doubtful relevance. Even contemporary textbooks in quantitative methods (for example, Griffith and Amrhein 1991; McGrew and Monroe 1993) discuss quite limited distance-based and areabased methods and do not consider the substantial and systematic advances in the statistical analysis of spatial point processes that have been made in the last twenty years.1 Given the particular area of application we consider here, the time is ripe for an assessment of the 'state of the art' in this field, though we focus on a sub-set of methods that have particular value in geographical epidemiology. Despite this emphasis, all of the methods we outline have applications in other areas of geographical inquiry.

Apart from well-understood shifts in disciplinary emphasis away from a perspective based on spatial analysis, there are perhaps two other reasons why spatial point pattern analysis has, until recently, been neglected in geography. The first (and more significant) reason is that the null hypothesis with which most of the early methods were concerned was rarely of real practical value. Typically, methods sought to establish departures from complete spatial randomness. Whilst this might prove a sensible benchmark in some cases, in others (such as examining the distribution of disease or the locations of retail outlets in urban areas) it is unlikely to prove illuminating. Although we shall make reference to the important concept of complete spatial randomness, we stress that the methods we outline go well beyond seeking solely to establish non-randomness. A second reason is simply the lack of availability of good software. While computer programs for nearest-neighbour or quadrat analysis were published, these generated purely textual output of statistical summaries and little or nothing in the way of maps or other graphical displays.

More recently, the statistical analysis of point patterns is attracting renewed interest, notably because of developments in geographical information systems (GIS). The proliferation of georeferenced databases, many of which generate data that may be treated as spatial point patterns, coupled with the need to infuse GIS with greater analytical functionality, have been major factors motivating the kind of work reported here and elsewhere (Gatrell and Rowlingson 1994). In particular, new tools have been developed for the analysis of point data; they are reviewed in the Appendix. These are now available as libraries that may be called from existing statistical programming environments, 'macros' that may be called from proprietary GIS packages, or functions within spatial analysis packages. They provide a variety of tools for the visualization, exploration and modelling of point data. In other words, they allow us simultaneously to view the point pattern, create new views of the pattern (for instance, showing variations in point density), explore structure in the data by estimating suitable summary functions and test hypotheses relating to the process that may have given rise to the observed event distribution.

Two final introductory remarks are in order. First, we observe that the use of spatial point pattern analysis in geographical epidemiology is hardly new, though some recent accounts in the epidemiology literature (Barreto 1993) seem to have discovered simple dot mapping as a useful technique!

Many accounts draw on the classic work of John Snow in Victorian London, linking the 'clustering' of cholera deaths around a pump in Soho to the probable source of infection – an example that appears in many introductory accounts of medical geography.

A wide range of analytical methods has been devised to handle spatial point patterns in epidemiology; we do not seek to review these comprehensively here, focusing instead on those methods we have found most useful in applied work. For example, one obvious omission in what follows is a discussion of Openshaw's 'geographical analysis machine' (Openshaw *et al.* 1987). This is now quite wellknown and is finding its way into texts on medical geography (Thomas 1993).

Secondly, we do not consider in detail how to obtain disease-incidence data. Suffice it to say that many epidemiological databases, particularly in Britain but also elsewhere, now contain a postcoded address that may be converted into a grid reference (Raper *et al.* 1992).

For example, in Britain, the direct link between unit postcodes and Ordnance Survey grid references with a resolution of 100 m (10 m in Scotland) means that one can readily produce mapped information on disease incidence as well as performing analyses of the point-event data, instead of aggregating these to areal units such as electoral wards.

The fact that one is not required to do such aggregation renders a point pattern approach attractive, since the results from any area-based analysis are dependent on the particular zoning system one uses. A priori it seems sensible to use methods that preserve the original continuous setting of the data. On the other hand, there are a number of questionable assumptions involved in accepting a unit postcode (referring, on average, to perhaps fifteen or so other households, though with some variation about this notional mean; *ibid.*) for it to be a sensible measure of location for the disease or an adequate reflection of exposure to risk factor(s). It suggests that the individuals forming the database of disease incidence are adequately represented by their address (strictly, postcode) at the time of diagnosis.

This assumes, quite naively, that people are immobile and ignores any possible exposure to environmental contamination (from whatever source) in the workplace or elsewhere. It further ignores the multitude of exposures to risk factors that may well have been picked up in earlier residential and occupational environments. However, we shall later see that in raised-incidence models we can begin to incorporate more meaningful covariates into the analysis and hence strive towards a richer interpretation and explanation of disease risk.

In the remainder of the paper, we first introduce some basic properties of spatial point processes and define some useful theoretical functions which may be used to characterize their behaviour.

We indicate how one would expect such functions to behave in a 'benchmark' theoretical situation and consider how to estimate such functions from an observed point pattern and how the results may be used to explore hypotheses of interest. We then look at the issue of spatial clustering in epidemiological data, followed by the extension to a spacetime context. Finally, we consider a modelling framework for assessing whether there is an elevated disease risk around a possible pollution source. In the Appendix we consider some software options for implementing the ideas developed here.

Concepts and Methods

Formally, a point pattern may be thought of as consisting of a set of locations ($s1$, $s2$, etc.) in a defined 'study region', R, at which 'events' of interest have been recorded. The use of the vector, si, referring to the location of the *ith* observed event, is simply a shorthand way of identifying the 'x' coordinate, $si1$, and the 'y' coordinate, $si2$, of an event. Use of the term 'event' has become standard in spatial point process analysis as a means of distinguishing the location of an observation from any other arbitrary location within the study region (Diggle 1983). The study region R might be a rectangular or complex polygonal region. Regardless of its shape, we must be aware of possible *edge effects* in the analysis, usually coping with these by either leaving a suitable *guard area* between the perimeter of the original study region and a sub-region within which analysis is performed, or by modifying the analytical tools to take account of boundary shape.

In the simplest case, our data set comprises solely the event locations. However, in some cases we may have additional information relating to the events which might have a bearing on the nature of analysis. For example, events may be of two different types (a *bivariate* point pattern), such as a set of individuals with a disease ('cases') and those without ('controls'). Alternatively, a continuous measure might be attached to each, an important instance being the time at which disease onset occurred among cases.

This gives rise to what is known as a *marked* point pattern. The simplest theoretical model for a spatial point pattern is that of *complete spatial randomness*, in which the events are distributed independently according to a uniform probability distribution over the region R. One important question that then arises is whether the observed events display any systematic spatial pattern or departure from randomness either in the direction of *clustering* or *regularity*.

However, the role of complete spatial randomness as such a benchmark is useful only in applications where departure from it is not obvious a priori. More interesting questions, especially in the human domain, include: Is observed clustering due mainly to natural background variation in the population from which events arise?

Over what spatial scale does any clustering occur? Are *clusters* merely a result of some obvious a priori heterogeneity in the region studied? Are they associated with proximity to other specific features of interest, such as transport arteries or possible point sources of pollution? Are events that aggregate in space also clustered in time? All these sorts of questions serve to take us beyond the simple detection of non-randomness and all are dealt with later.

The solutions to many mathematical questions, both pure and applied, rely on the ability of the investigator to uncover a pattern. In basic terms, Point Pattern Analysis is an investigation focused on finding patterns in data comprised of points in a spatial region. One common application of Point Pattern Analysis is epidemiology. The medical community is often interested in the spread of infectious disease such as: SARS, chicken pox, and West Nile virus among others. It is possible to identify pattern to the spread of infection then this might lead to an understanding of how the spread of an illness is related to social behaviour, environmental factors, genetic susceptibility, or many other health care factors.

In general, a spatial data set takes the form: X= $\{x_k \mid x_k \varepsilon R^{m,} m \in N\}$. However, it is possible for the data to contain spatial location plus additional information. For example, earthquake data typically gives the location of earthquakes along a fault line and will often have the size and the time of each earthquake. Data that contains spatial data plus additional information is often referred to as *marked spatial data*. In our analysis, we will be concerned with only the spatial information and we will disregard any additional information associated with the data. Moreover, the examples we will work with are limited to two-dimensional data.

Our interest will lie in quantifying the dispersion of objects within a confined geographical area. We try to understand the interaction of pattern and process and use point pattern analysis as a mechanism for detecting patterns associated as compared to random processes. The random process that will serve for our comparison will be the homogenous Poisson process, which will be described in more detail in section 2.

D.J Gerrard describes an investigation of a 19.6 acre square plot in Lansing Woods, Michigan . This data includes hickories, maples and oaks grown on a square plot. The data for hickories is given in Cartesian coordinates, that is, (x_i, y_i) form, where x_i *and* $y_i \in R$. Also, the points are plotted on a unit square region.Our main goal of Point Pattern Analysis is to find out whether the distribution of the hickory trees is random, clustered or regularly dispersed.

The kind of pattern involved would further our understanding of the behaviour of the hickory trees and thus can be of great use to ecologists and biologists. For example, if the pattern is clustered, the biologists may conclude that natural factors encourage the hickories to cohabitate and promote tree growth.

There are several methods and algorithms that endeavor to describe pattern for a collection of points. The most common methods discovered for spatial pattern analysis are as follows:

1. Quadrant Count Method
2. Kernel Density Estimation (K means)
3. Nearest Neighbor Distance

a. G function

b. F function

c. K function.

The above list of techniques for Point Pattern Analysis is among the most popular and best established mathematical and statistical methods used in the literature [1,2,4,5]. Since Point Pattern Analysis can take several forms and can be applied in a variety of settings we will present a list of criteria in order to determine if a data set is suitable for our Point Pattern Analysis.

The criteria we will use to determine if a data set is appropriate for our type of point patter analysis is given by the following:

- Spatial data must be mapped on a plane; both latitude and longitude coordinates are needed.
- The study area must be selected and determined prior to the analysis.
- Point data should not be a selected sample, but rather the entire set of data to be analyzed.

- There should be 1-1 correspondence between objects in study area and events in pattern.
- Points must be true incidents with real spatial coordinates.

Since this is an introductory venture into the subject of Point Pattern Analysis we have selected the Quadrant Count Method for our analysis. While the other techniques listed in this section can be more descriptive and more accurate it is also true that many of these other methods are more complicated and difficult to implement. We have found that Quadrant Count Analysis is relatively easy method to implement and it has provided several opportunities to apply basic mathematical and statistical concepts.

Quadrant Count Method

The Quadrant Count Method can be described simply as partitioning the data set into *n* equal sized sub regions; we will call these sub regions quadrants. In each quadrant we will be counting the number of events that occur and it is the distribution of quadrant counts that will serve as our indicator of pattern. The choice of the quadrant size can greatly affect our analysis, where large quadrants produce a coarse description of the pattern. If the quadrant size is too small then many quadrants may contain only one event or they might not contain any events at all. We will use the rule of thumb for the area of a square is twice the expected frequency of points in a random distribution (i.e., $2\frac{Area}{n}$), where n is the number of points in the sample size. After partitioning the data set into quadrants, the frequency distribution of the number of points per quadrant has been constructed. The Mean and Variance of the sample are then computed to calculate the Variance-to-Mean Ratio (VTMR). The following is the way we will interpret the VTMR of a sample:

- If VTMR>1, the pattern is clustered. This implies that the data set has one or more groups of points in clusters and large areas of maps without points.

The region might look like Figure 1:

Figure 1 A Clustered pattern

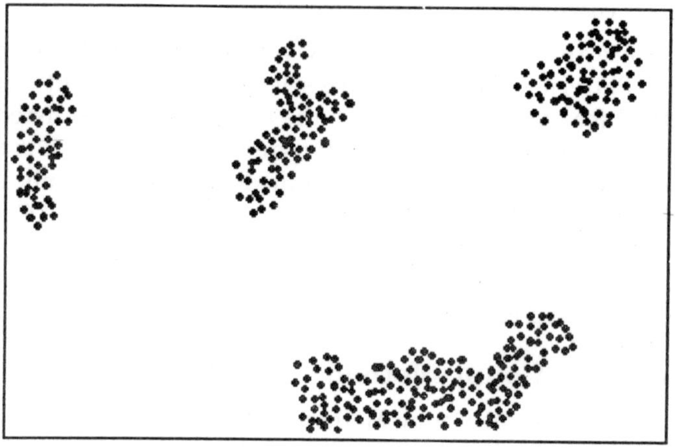

- If VTMR<1, the pattern is regularly dispersed implying the events are distributed more or less regularly over the region. A regularly dispersed area might look like Figure 2:

Figure 2 Regularly Dispersed Pattern

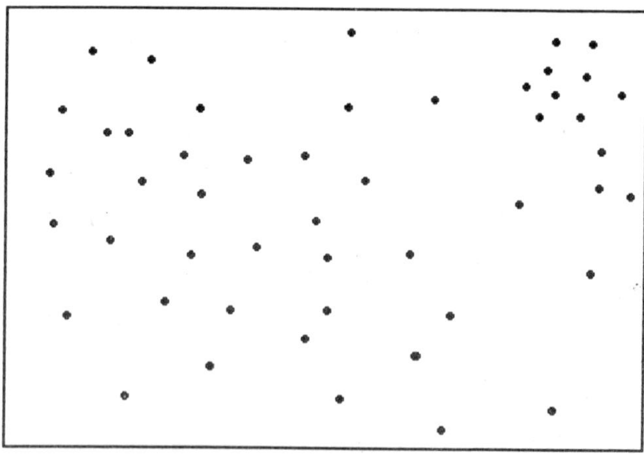

- If VTMR=1, the pattern is random. This implies the data set has no dominant trend towards clustering or dispersion. A random pattern may look like Figure 3:

Figure 3 Random Pattern

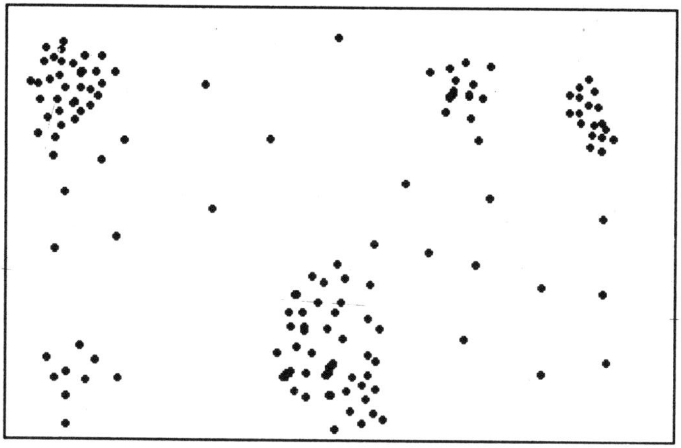

The random model that will serve as our standard of comparison is the *Complete Spatial Randomness* (CSR) model [1,3]. The CSR model has two basic characteristics:

(1) The number of events in any planar region A is with area $|A|$ follows a Poisson distribution with mean ë $|A|$.

(2) Given there are n events in A, those events are independent and form a random sample from a uniform distribution on A.

The constant ë is the intensity, or the mean number of events per unit area. Also, by (1), the intensity of events does not vary over the plane. According to (2), CSR also implies, the events are independent of each other and there is no interaction between them.

The mathematical construct that we will use to simulate a CSR model is the homogenous Poisson process. The Poisson process is suitably defined by the following postulates:

(a) If ë>0, and any finite planar region A, *N (A)*, follows a Poisson distribution with mean ë $|A|$.

(b) Given *N (A)=n*, the *n* events in A form an independent random sample from the uniform distribution on A. (In our case, n is the number of trees in the region)

As stated above, the CSR corresponds to the homogenous Poisson Distribution. Please recall that the Poisson distribution can be used in place of the Binomial distribution in the case of very large samples. In a binomial distribution, all eligible phenomena are studied, whereas in the Poisson distribution only the cases with a particular outcome are studied. The Poisson distribution can also be used to study how "events" are distributed on the level of a population. If having one "event" has no influence on the chance of having another accident, the "event" is put back into the population immediately after an "event"; people may have one, two, three, or more events during a certain period of time. One assumption in this application of the Poisson distribution is that the chance of having an event is randomly distributed: every individual has an equal chance. Mathematically, this is expressed in the fact that for a Poisson distribution, the variance of the sample is equal to its mean. Hence, in the QCM, the VTMR of a random sample is equal to 1, since the variance is equal to its mean.

How to Apply Quadrant Count Analysis

To explain the application of QCA in detail, a small data set is plotted on a square region. The region is then divided into equally sized quadrants (squares). This is demonstrated in the following figure. For the sake of simplicity, we have chosen to divide the region into only four quadrants as illustrated in Figure 4.

Figure 4 Dividing a region into quadrants

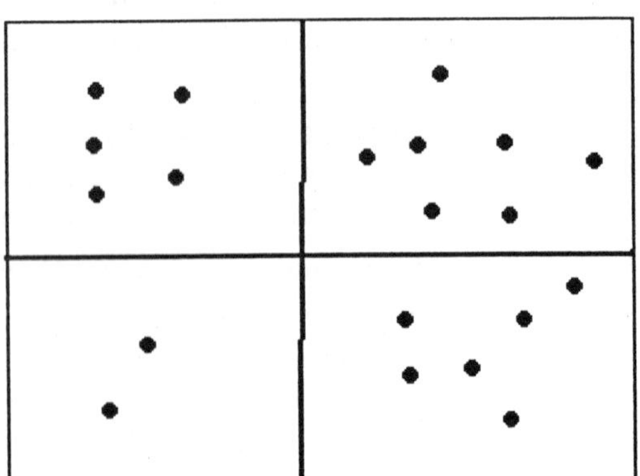

The Mean of the sample can then be calculated as:

$$Mean = \frac{No.\ of\ pts. in the region}{No. of\ quadrants} = \frac{20}{4} = 5.$$

Let x_i be the frequency of points in each quadrant. Then Variance of the sample can the be calculated as

$$Variance = \frac{\sum x_i^2 - \frac{(\sum x_i)^2}{n}}{n-1} = \frac{2^2 + 5^2 + 6^2 + 7^2 - \frac{(20)^2}{4}}{4-1} = 4.5$$

The Variance to Mean Ratio or VTMR is calculated as

$$VTMR = \frac{Variance}{Mean} = \frac{4.5}{5} = 0.9.$$

According to QCM, since VTMR for this example is less than 1, our simple data set is classified as regularly dispersed. Thus the points tend to repel each other and are thought to spread evenly throughout the region.

Using QCM to Analyze the Hickories

To apply Quadrant Count Analysis to our data, we used a code written in C++. We experimented with different quadrant sizes, using $4^m, m \in N$ as our number of quadrants. This choice of number of quadrants made it easier to experiment with different number of quadrants. Table 1 displays the results generated with different quadrant sizes.

Table 1 VTMR for different quadrant sizes

N	Grid Size	Mean	Variance	VTMR
1	2*2	175.75	3808.33	21.669
2	4*4	43.9375	529	12.0398
3	8*8	10.9844	55.6032	5.06202
4	16*16	2.74609	6.52941	2.37771
5	18*18	2.16975	4.29721	1.98051

6	19*19	1.94837	3.57778	1.83724
7	32*32	0.686523	0.939394	1.36833
8	64*64	0.171631	0.188767	1.09884
9	128*128	0.0429077	0.0443753	1.0342
10	256*256	0.0107269	0.0109865	1.0242

As we test different quadrant sizes, we notice the smaller number of quadrants corresponds to larger variance. But as we divide the region into smaller quadrants, the variance starts decreasing to one.

We chose to divide the hickories into 256 quadrants (16*16 grid) as our experimenting size. We select this number since it is the power of 4 that is to the rule of thumb: number of quadrants=. For 256 quadrants the VTMR was found to be approximately 2.37771.

According to the QCM, a VTMR that is greater than 1 implies that the hickories must be clustered. However, it is possible of a data set that is randomly generated would also have a VTMR that is greater than one when counting over 256 quadrants. How confident can we be that our 2.38 VTMR does indeed correspond to the clustering pattern of our hickories? The answer lies in using the homogeneous Poisson process to simulate random data for the purpose of comparison.

Simulations

To be more confident about our hickories being clustered, we generate 500 data sets of 703 random points each. Their VTMR is then calculated and a histogram of the 500 VTMR is built.

We will use the empirical p-value of our observed value of VTMR (2.37771) in correspondence with our random simulations. Please recall the empirical p-value is the percentage of the VTMR from the random samples that are greater than or equal to our observed VTMR. Hence, a small the empirical p-value implies that we can be more confident that our data set is clustered. It turns out that none of the VTMR of the random samples are equal to or greater than our calculated VTMR. Therefore, our Empirical p-value is 0 and we can be highly confident that our data set of hickories is clustered.

To illustrate the utility of the empirical p-value consider the hypothetical case that we had a set of tree data that had a VTMR is very close to 1, say 1.01. The Quadrant Count Method would classify this data set as clustered,. If we utilize the same collection of 500 random data sets and calculate the empirical p-value that corresponds to an observed VTMR of 1.01 then our empirical p-value is approximately.58. This value implies more than half the random simulations have a VTMR greater than or equal to 1.01. Thus, we cannot be confident that data set with a VTMR of 1.01 is in reality clustered.

8

Inference for Spatial Data

Spatial changes within an environment are typically a result of interaction— actions and events— occurring within. Reasoning about such changes, when dealt with formally within the context of qualitative spatial calculi and logics of action and change, poses several difficulties along multiple dimensions: (a) phenomenal requirements stemming from the dynamic nature of the spatial system (e.g., appearing and disappearing objects), (b) reasoning requirements (e.g., abductive explanation), (c) domain-independent or epistemological (e.g., persistence, ramification), and (d) aspects concerning the need to satisfy the intrinsic (axiomatic) properties of the spatial calculi (e.g., compositional consistency) being modelled. This paper, encompassing the phenomenal and reasoning aspects in (a) and (b) respectively, presents some instances that demonstrate the role of commonsense reasoning and the non-monotonic inference patterns it necessitates whilst representing and reasoning about dynamic spatial systems in general.

Motivation

Dynamic Spatial Systems (DSS) are systems where spatial configurations, denoted by sets of qualitative spatial relations, undergo transformations as a result of actions and events occurring within the environment [Bhatt and Loke, 2008]. The DSS approach is applicable in a wide-range of application domains as diverse as cognitive robotics, diagrammatic reasoning, architecture design, geographical information systems and even the new generation of ambient intelligence systems involving behaviour or activity monitoring. From the viewpoint of such applications, the basic functionality required from the DSS approach remains the same, namely, the capability to serve either a predictive (i.e., projection, planning) or an explanatory (e.g., causal explanation) function in the context of high-level qualitative models of space and spatial change. This in turn requires that change in general and spatial change in specific, and its relationship to action, events and other aspects such as causality be taken seriously.

Existing qualitative spatial modelling techniques have primarily remained focused on reasoning with static spatial configurations. In general, research in the qualitative spatial reasoning domain has remained focused on the representational aspects of spatial information conceptualization and the construction of efficient computational apparatus for reasoning over those by the application of constraint-based techniques [Cohn and Renz, 2007, Renz and Nebel, 2007].

For instance, given a qualitative description of a spatial scene, it is possible to check for its consistency along arbitrary spatial domains (e.g., topology, orientation and so forth) in an efficient manner by considering the general properties of a qualitative calculus [Ligozat and Renz, 2004].

However, for applications such as the ones aforementioned, these methods require a realistic interpretation, such as the one provided by the stated DSS perspective, where sets of spatial relations undergo change as a result of named occurrences in the environment, or broadly, reasoning about space and reasoning about actions and change are consilidate into a 'Reasoning about Space, Actions and Change' (RSAC) paradigm [Bhatt, 2009]. Consequently, the formal embedding of arbitrary spatial calculi – whilst preserving their high-level axiomatic semantics and if necessary, their low-level algebraic properties too – has to be investigated from the viewpoint of formalisms that deal with action and change in general.

In this paper, we illustrate the utility of commonsense reasoning, and the non-monotonicity it entails, toward realising the suggested embedding of spatial calculi within general formalisms of action and change. Note that the embedding per se is extensive, and not the object of this paper. Rather, we solely focus on some commonsense inference patterns that occur whilst achieving the said embedding. These patterns pertain to the following aspects:

AI existential consistency of complete spatial situation descriptions given that fact that the domain of discourse of primitive spatial entities may be incompletely known, i.e., unknown objects may have appeared or known objects may have either temporarily disappeared or may have permanently ceased to exist AII modelling causal explanation tasks, where given a set of temporally-ordered observations, the objective is to derive an explanation in terms of the (spatial and nonspatial) actions and events that may have caused the observations.

Here, modelling causal explanation abductively necesssitates the use of a circumscriptive non monotonic approach. In comparision to the other epistemological and intrinsic spatial calculi related commonsense patterns, which are excluded from this paper, the aspects in (AI–AII) are extrinsic to the process of embedding and concern phenomenal aspects (section 3) of dynamic spatial systems and the computational or reasoning requirements (section 4) expected from an operationalization of the DSS perspective.

Qualitative Spatial Primitives

The objective of this paper is to intuitively present the nature of commonsense reasoning as relevant to aspects (AI–AII; section 1). As such, we do not go into the details of the formal axiomatisation of a theory of change or the details pertaining to the constitution of a qualitative spatial calculus. However, a basic overview of the ontological setup is needed to make the paper self-contained.

The spatial ontology that is required depends on the nature of the spatial calculus that is being modeled. In general, spatial calculi can be classified into two groups: topological and positional calculi. When a topological calculus such as the Region Connection Calculus (RCC) [Randell et al., 1992] is being modeled, the primitive entities are spatially extended and could possibly even be 4D spatio-temporal histories (e.g., in a domain involving the analyses of motion-patterns). Alternately, within a dynamic domain involving translational motion in a plane, a point-based (e.g., Double Cross Calculus [Freksa, 1992], OPRAm [Moratz, 2006]) or line-segment based (e.g., Dipole Calculus [Schlieder, 1995, Moratz et al., 2000]) abstraction with orientation calculi suffices. This fragment consists of eight relations: disconnected (dc), externally connected (ec), partial overlap (po), equal (eq), tangential proper-part (tpp) and non-tangential proper-part (ntpp), and the inverse of the latter two tpp-1 and ntpp-1. Similarly, illustrates one primitive relationship for the Oriented Point Relation Algebra (OPRA) [Moratz, 2006], which is a spatial calculus consisting of oriented points (i.e., points with a direction parameter) as primitive entities.

The granularity parameter m determines the number of angular sectors, i.e., the number of base relations. Applying a granularity of m = 2 results in 4 planar and 4 linear regions, numbered from 0 to 7, where region 0 coincides with the orientation of the point. The family of OPRAm calculi are designed for reasoning about the relative orientation relations between oriented points and are well-suited for dealing with objects that have an intrinsic front or move in a particular direction. Spatial Situation Descriptions Spatial situation descriptions consist of a complete n-clique graph for a domain of n objects. Further, there is one such clique for every type of spatial domain (e.g., topology, orientation) that is modelled. Precisely, for a spatial scene description with n domain objects and k spatial calculi being modeled, the scene description involving n objects requires a complete n-clique specification with [n(n ÀÛÜ 1)=2] spatial relationships for each of the respective calculi. Given such spatial scene descriptions, the following notion of existential consistency is definable:

A spatial scene description is E-Consistent, i.e., existentially consistent, if there exists at least one spatial relationship of any spatial domain (i.e., topology, orientation etc) that every existing spatial object participates in with other existing object(s).

From the viewpoint of planning and explanation tasks, E-Consistency is necessary and useful toward maintaining the consistency of spatial scene descriptions given the fact that appearance of new objects and disappearance of existing ones may have occurred within the system. In the context of such phenomenal requirements, the significance and use of E-Consistency from Definition 2.1 is further elaborated on in section 3.

Phenomenal Commonsense: Appearance and Disappearance of Objects

Appearance of new objects and disappearance of existing ones, either abruptly or explicitly formulated in the domain theory, is characteristic of non-trivial dynamic spatial systems. In robotic applications, it is necessary to introduce new objects into the model, since it is unlikely that a complete description of the robot's environment is either specifiable or even available.

Similarly, it is also typical for a mobile robot operating in a dynamic environment, with limited perceptual or sensory capability, to lose track of certain objects because of issues such as noisy sensors or a limited field-of-vision. As an example, consider a 'delivery scenario' in which a vehicle/robot is assigned the task of delivering 'object(s)' from one 'way-station' to another. In the initial situation description, the domain consists of a finite number of 'way-stations' and deliverable 'objects'.

However, the scheduling of new objects for delivery in future situations will involve introducing new 'objects' into the domain theory.

For example, an external event1 such as 'schedule delivery(new load; loc1; loc3)' introduces a new object, namely 'new load', into the domain. Appearance and disappearance events involving the modification of the domain of discourse are not unique to applications in robotics. Even within the projected next-generation of event-based and temporal geographic information systems, appearance and disappearance events are regarded to be an important typological element for the modelling of dynamic geospatial processes [Claramunt and Th´eriault, 1995, Worboys, 2005].

For instance, Claramunt and Th´eriault [1995] identify the basic processes used to define a set of low-order spatio-temporal events which, among other things, include appearance and disappearance events as fundamental. Similarly, toward event-based models of dynamic geographic phenomena, Worboys [2005] suggests the use of appearance and disappearance events at least in so far as single object behaviours are concerned. We regard that such phenomena, being intrinsic to a typical dynamic spatial system, merit systematic treatment.

Maintaining and Propagating Existential Facts

The case of disappearance is not problematic, however, for the case of appearance and re-appearance, some questions that need to be addressed include: _ what is the spatial relationship (topological, directional etc) of the newly appearing object with other existing objects? _ given the fact that a newly appearing object is, from a model-theoretic viewpoint, unknown in the past, how to make it 'known' and 'not exist' in the past?

A Planner's Encounter with Complexity

Spatial planning is about dealing with our 'everyday' environment. In A Planner's Encounter with Complexity we present various understandings of complexity and how the environment is considered accordingly.

One of these considerations is the environment as subject to processes of continuous change, being either progressive or destructive, evolving non-linearly and alternating between stable and dynamic periods. If the environment that is subject to change is adaptive, self-organizing, robust and flexible in relation to this change, a process of evolution and co-evolution can be expected.

This understanding of an evolving environment is not mainstream to every planner. However, in A Planner's Encounter with Complexity, we argue that environments confronted with discontinuous, non-linear evolving processes might be more real than the idea that an environment is simply a planner's creation. Above all, we argue that recognizing the 'complexity' of our environment offers an entirely new perspective on our world and our environment, on planning theory and practice, and on the raison d'être of the planners that we are.

Entity-relationship Modelling of Spatial Data for Geographic Information Systems

GIS design (planning, design and implementing a geographic information system) consists of several activities: feasibility analysis, requirements determination, conceptual and detailed database design, and hardware and software selection. Over the past several years, GIS analysts have been interested in, and have used, system design techniques adopted from software engineering, including the software life-cycle model which defines the above mentioned system design tasks (Boehm, 1981, 36). Specific GIS life-cycle models have been developed for GIS by Calkins (1982) and Tomlinson (1994).

GIS design models, which describe the implementation procedures for the life-cycle model, outline the basic steps of the design process at a fairly abstract level (Calkins, 1972; Calkins, 1982). While useful as basic guides for GIS designers, the present models are deficient in that they describe only high-level activities. Additionally, these guides usually do not describe any methods for completing the indicated design tasks. One aspect of these models that has not been given sufficient additional definition is the conceptual and detailed design procedures for geographic databases. An extensive literature on general database design techniques exists, however this body of knowledge has yet to be adapted for use in designing geographic databases.

GIS design methodologies currently in use do not treat the database design problem in detail. Often, data of interest are simply listed in tabular form on the assumption that using a commercial GIS format obviates the need for any further effort toward database design. More enlightened GIS design approaches link the data needed in the GIS to applications and GIS procedures. These approaches, however, do not offer any assistance in either the conceptual or logical design of the GIS database. Errors in database design can still occur and can be very costly. More attention needs to be paid to geographic database design. Specific tools reflecting the special characteristics of spatial data need to be developed to support the database design portion of the GIS design process. The building of the database for a GIS is frequently the largest single cost item, consuming as much as 80% of the total GIS project cost (Dickinson and Calkins, 1989). In this environment, it is well worth developing enhanced tools for supporting the GIS database design activities.

Of the many specific database design techniques developed, the entity-relationship (E-R) modelling technique developed by Chen (Chen, 1976), has gained popularity and been extremely effective over a wide range of application areas. This paper presents a proposed extension to the basic E-R modelling technique specifically for describing geographic data for use in the process of designing GIS database.

Conceptual Modelling and Data Base Design

Database design is the information system planning activity where the contents of the intended database are identified and described. Data base design is usually divided into three major activities (Elmasri and Navathe, 1994, 41):

(1) Conceptual data modelling: identify data content and describe data at an abstract, or conceptual, level;

(2) Logical database design: translation of the conceptual database design into the data model of a specific software system; and

(3) Physical design: representation of the data model in the schema of the software.

For most GIS implementations, a commercial GIS software package is used, often in conjunction with a commercial database system. In these instances, the basic structure of the logical schema (e.g., a relational data schema) and the entire physical schema are already predetermined. The task of the GIS designer is to prepare a conceptual schema that properly describes the entire GIS database and is suitable for translation into the logical schema of the proposed GIS and database software.

To achieve a good database design, and thereby the desired GIS, the conceptual data model must be complete, i.e., contain all data needed to meet the system's objectives, and must be directly translatable into the logical and physical database schema. Data identification and description includes defining the objects (entities), the relationships between the objects and the attributes of objects (or relationships) that will be represented in the database. The data description activity also includes assembly of information about the data objects, i.e., metadata (definition, data type, valid values, data quality, etc.).

The purpose of the conceptual data modelling process is to prepare an unambiguous and rigorous description of the data to be included in the database in a form that: 1) is understandable by the proposed users of the database or system; and 2) is sufficiently structured for a programmer or analyst to design the data files and implement data processing routines to operate on the data. The emphasis is on: 1) communication between the user and the programmer/analyst; and 2) review and verification of the data model and database design by both user and analyst. In the past, in a typical GIS design activity, users/analysts have tended to describe their data needs in general, somewhat vague, terms, such as a simple list. The programmer/analyst needs precise information about the data to set-up the database and necessary computer processes. Descriptions of data and algorithms using normal language (such as English) are not usually adequate for implementing a system.

Thus, tools allowing greater precision in description are needed for the task of conceptual data modelling and database design. Such tools provide a means to identify and describe the intended database in terms that facilitate two critical activities: user verification and detailed database design. Conceptual modelling is the representation of the functional application requirements and information system components at a abstract level, i.e., a description of what is to be included into an information system rather than how the intended information system will work. "The quality of a conceptual schema (model) and ultimately that of the information system depends largely on the ability of a developer to extract and understand knowledge about the modeled domain" (Loucopoulos and Zicari (1992). Further, "a conceptual schema of a concrete system is called a conceptualization and represents the basic specification mechanism during requirements analysis. A symbolic representation of a conceptual system is called a representation;

for example, the Entity-Relationship Diagram is a representation of the Entity-Relationship Model".

A tool to support conceptual data modelling should have the above characteristics in addition to the ability to ensure completeness of description and the ability to support one-tone mapping of the conceptual model into the logical database design. For an extended discussion of conceptual modelling and information system design, the reader is referred to Loucopoulos and Zicari, 1992).

Tools for Conceptual Database Modelling

Various tools to support the conceptual database modelling activity have been developed. One widely accepted and used technique is the entity-relationship modelling technique developed by Chen (Chen, 1976). The entity-relationship (E-R) technique has been applied to many disciplines and has been revised and extended by many researchers to meet a variety of specialized needs. The E-R technique is a graphical method of representing objects (or entities) of a database, all important relationships between the entities, and all attributes of either entities or relationships which must be captured in the database. The set of rules controls the definition of entities, relationships, and attributes and the manner in which they are portrayed in diagrammatic form. All items are names and additional information is appended to the diagram indicating the nature of each relationship, i.e. the cardinality of each relationship (one-to-one, one-to-many, or many-many).

Basic Entity-Relationship Modelling

The basic entity-relationship modelling approach is based on describing data in terms of the three parts noted above (Chen 1976):

- Entities
- Relationships between entitles
- Attributes of entitles or relationships.

Each component has a graphic symbol and there exists a set of rules for building a graph (i.e., an ER model) of a database using the three basic symbols. Entities are represented as *rectangles,* relationships as *diamonds* and attributes as *ellipses.* The normal relationships included in a E-R model are basically those of:

1. Belonging to;
2. Set and subset relationships;
3. Parent-child relationships; and
4. Component parts of an object.

The implementation rules for identifying entities, relationships, and attributes include an English language sentence structure analogy where the nouns of a descriptive sentence identify entities, verbs identify relationships, and adjectives identify attributes. These rules have been defined by Chen (1983) as follows:

Rule 1: A common noun (such as person, chair), in English corresponds to an entity type on an E-R diagram.

Rule 2: A transitive verb in English corresponds to a relationship type in an E-R diagram.

Rule 3: An adjective in English corresponds to an attribute of an entity in an E-R diagram.

English statement: A person may own a car and may belong to a political party. Analysis: "person," "car," and "political party" are nouns and therefore correspond to entity types. ... "own" and "belong to" are transitive verbs (or verb phases) and therefore define relationships.

The process of constructing an E-R diagram uncovers many inconsistencies or contradictions in the definition of entities, relationships, and attributes. Many of these are resolved as the initial E-R diagram is constructed while others are resolved by performing a series of transformations on the diagram after its initial construction.

For a discussion of E-R transformations see Jajodia and Ng (1983). The final E-R diagram should be totally free from definitional inconsistencies and contradictions. If properly constructed, an E-R diagram can be directly converted to the logical and physical database schema of the relational, hierarchical or network type database for implementation.

Geographic Data Models

The data models in most contemporary GISs are still based on the cartographic (or spatial) data object view. Other data models have begun to evolve, but are still very limited. Current and potential geographic data models include:

- The cartographic data model: points, lines and polygons (topologically encoded) with one, or only a few, attached attributes, such as a land use layer represented as polygons with associated land used code.
- Extended attribute geographic data mode: geometric objects as above but with many attributes, such as census tract data sets;
- Conceptual object/spatial data model: explicit recognition of user defined objects, zero or more associated spatial objects, and sets of attributes for reach defined object (example: user objects of land parcel, building, and occupant, each having its own set of attributes but with different associated spatial objects: polygon for land parcel, footprint for building, and no spatial object of occupant).
- Conceptual objects/complex spatial objects: multiple objects and multiple associated spatial objects (example: a street network with street segments having spatial representations of both line and polygon type and street intersections having spatial representations of both point and polygon type). Current GIS are based on the cartographic and extended attribute data models. Data modelling tools supporting GIS design will need to accommodate all the above defined data modelling cases. The remainder of this paper describes how an extension to the E-R modelling technique can provide the necessary modelling tool for GIS database design.

Representation of Spatial Relationships

The E-R methodology rules, particularly the sentence verb rule, can be applied to the identification of the spatial relationships found in a geographic information system. Table 2 includes the common set of spatial relationships of interest in a GIS. In a geographic information system, spatial relationships are implemented into one of three ways:

- topological encoding;
- x,y coordinates and accompanying spatial operation; and
- the one of the traditional database relationships previously identified.

Connectivity and contiguity are implemented through topology: the link-node structure for connectivity through networks and the arc-polygon structure for contiguity. Containment and proximity are implemented through x,y coordinates and related spatial operations: containment is determined using the point-, line-, and polygon-on-polygon overlay spatial operation and proximity is determined by calculating the coordinate distance between two or more x,y coordinate locations. The spatial relationship of coincidence may be complete coincidence or partial coincidence. The polygon-on-polygon overlay operation in ARC/INFO™ calculates partial coincident of polygons in two different coverages. The System 9™ Geographic Information System recognizes coincident features through a "shared primitive" concept (the geometry of a point or line is stored only once and then referenced by all features sharing that piece of geometry). Future versions of commercial GISs will likely implement coincident features through either the "belonging to" database relationship or through x,y coordinates and related spatial operations, whichever is more efficient within the particular GIS.

Modelling Geographic Data with E-R Techniques

Previous use of E-R modelling techniques for geographic database design have been reported by Calkins and Marble (1985), Bedard and Paquette (1988), Wang and Newkirk (1988), and Armstrong and Densham (1990). Calkins and Marble (1985) demonstrated the use of the basic ER) methodology in the design of a cartographic database where there was no special provision for the geographic entities of points, lines, or polygons in the integrated E-R diagram. However, Calkins and Marble (1985, 118) did demonstrate that the E-R diagram of the cartographic database could be transformed into an E-R representation based solely on the spatial entities. Bedard and Paquette (1988) and Wang and Newkirk (1988) include the spatial entities in their E-R diagrams by sing the "ISA" relationship where each entity is defined normally and also as a spatial entity.

Armstrong and Densham (1990, 16) defined an extended network representation for spatial decision support systems (SDSS).

All of the above efforts are limited and insufficient for GIS database design in that the identification of the spatial entity corresponding to the real world object being modeled is either missing, presented as a separate or additional entity, or is otherwise redundantly represented in the E-R diagram. The resulting E-R diagrams are less easily understood, are more complex than need be, and are less easily manipulated (or transformed) to remove errors or inconsistencies.

Additionally, and most importantly, there is no direct mapping (1:1) to a logical database schema, a primary criterion for a useful database design tool. Thus, these E-R extensions do not adequately meet the communications or operability goals for a database design tool.

Additionally, and more importantly, neither the basic E-R modelling technique nor any of the extensions developed represent the full extent of the spatial relationships that exist in a geographic information system and its associated database. Her it is necessary to recognize a significant difference between a GIS and other types of information systems, particularly business oriented information systems.

In a typical business application, all relationships between and among data objects can be explicitly represented in the database and the associated software is usually oriented to query and report generation functions. In a GIS, the set of relationships between and among data objects is represented, in part, explicitly in the database and, in part, is implemented through various software functions. For example, there is a spatial relationship between two map layers of the same area or location (spatial coincidence). However, the GIS database usually does not explicitly represent this relationship. The relationship is derived through the use of the topological overlay spatial operation, which is a software function. It is the latter case (a data relationship that will be implemented through software) that neither the basic E-R modelling methodology, nor any proposed extension, can adequately describe.

Further, additional demands on the data modelling tool will come from the continuing evolution of GISs. It is now considered necessary for such a system to accommodate multiple spatial representations of a single entity and multiple temporal representations of a single entity.

Finally, an adequate database design tool for geographic databases needs a structure to represent the database in compact form, as a geographic database usually has a large number of entities and relationships that yield large and complex diagrams. To meet the objective of communication and understandability the graphic tool must be as compact as possible.

Modelling a Geographic Database

Modelling a geographic database using the E-R approach requires an expanded or extended concept for:

- Entity identification and definition; and
- Relationship types and alternate representational forms for spatial relationships.

There are three considerations in the identification and definition of entities in a geographic database:

Correct Identification and Definition of Entities

Entities in a geographic database are defined as either discrete objects (e.g., a building, a bridge, a household, a business, etc.) or as an abstract object defined in terms of the space it occupies (e.g., a land parcel, a timber stand, a wetland, a soil type, a contour, etc.). In each of these cases we are dealing with entities in the sense of "things" which will have attributes and which will have spatial relationships between themselves. These "things" can be thought of as "regular" entities.

Defining a Corresponding Spatial Entity for Each "Regular" Entity

A corresponding spatial entity will be one of the spatial data types normally handled in a GIS, e.g., a point, line, area, volumetric unit, etc. The important distinction here is that we have a single entity, its spatial representation and a set of attributes; we do not have two separate objects. A limited and simple set of spatial entitles may be used, or alternatively, depending on the anticipated complexity of the implemented geographic information system, an expanded set of spatial entities may be appropriate. The corresponding spatial entity for the regular entity may be implied in the definition of the regular entity, such as abstract entities like a wetland where the spatial entity would normally be a polygon, or a contour where the spatial entity would be a line. Other regular entitles may have a less obvious corresponding spatial entity. Depending on the GIS requirements, the cartographic display needs, the implicit map scale of the database and other factors, an entity may be reasonably represented by one of several corresponding spatial entities. For example, a city in a small-scale database could have a point as its corresponding spatial entity, while the same city would have a polygon as its corresponding spatial entity in a large-scale geographic database.

Multipurpose (or corporate) geographic databases may need to accommodate multiple corresponding spatial entities for some of the regular entities included in the GIS. For example, the representation of an urban street system may require that each street segment (the length of street between two intersecting streets) be held in the GIS as both a single-line street network to support address geocoding, network based transportation modelling, etc., and as a double-line (or polygon) street segment for cartographic display, or to be able to locate other entities within the street segment (such as a water line), etc. In each of these instances the "regular" entity is the street segment, although each instance may have a different set of attributes and different corresponding spatial entities. Also, there may be a need to explicitly recognize multiple temporal instances of regular entities. The simple case of multiple temporal instances will be where the corresponding spatial entity remains the same, however, future GISs will, in all likelihood, have to deal with multiple temporal instances where the corresponding spatial entity changes over time.

In a manner similar to the expansion of the concept of entities, the concept of relationship also needs to be expanded. In addition to the regular, or normal, types of relationship that can exist between entities, a set of spatial relationships need to be defined and included in the E-R model. Modelling a geographic database using the E-R framework requires additional notation to represent the five spatial relationships and the manner in which they will be implemented. This is accomplished in the E-R model extension for spatial data in two ways: 1) a change in the entity symbol providing for inclusion of information about the corresponding spatial object and associated spatial codes; and 2) defining additional symbols for representing spatial relationships.

Three symbols are defined to represent entities: entity (simple); entity (multiple spatial representations); and entity (multiple time periods). The internal structure of the entity symbol contains the name of the entity and additional information indicating the corresponding spatial entity (point, line or polygon), a code indicating topology, and a code indicating encoding of the spatial entity by coordinates. The coordinate code is, at the present time, redundant in that all contemporary GISs represent spatial entities with x,y coordinates. However, it is possible that future geographic databases may include spatial entities where coordinates are not needed. Similarly, topological encoding is normally of only one type and can, for the present, be indicated by a simple code. However, different spatial topologies have been defined and may require different implementations in a GIS (Armstrong and Densham, 1990). In the future, the topology code may be expanded to represent a specific topologic structure particular to a GIS application.

The spatial relationships are defined by three relationship symbols. The traditional diamond symbol can be used for normal database relationships. An elongated hexagon and a double elongated hexagon, are defined to represent spatial relationships. The elongated hexagon represents spatial relationships defined through topology (connectivity and contiguity) and the double elongated hexagon represents spatial relationships defined through x,y coordinates and related spatial operations (coincidence, containment and proximity). The appropriate "verbs" to include in the hexagonal symbols are the descriptors of the spatial relationships. The spatial operation will be implicitly defined by the relationship symbol (double hexagon), the spatial entity and the topology code. For example, a spatial relationship named "coincident" between entities named "wetlands" and "soils," both of which carry topologic codes and x,y coordinates, indicates the spatial operation of topological overlay. If this does not sufficiently define the spatial operation needed, the name of the spatial operation can be used to describe the relationship, such as shortest path, point-in-polygon, radial search, etc.

Conceptual Model of a Spatial Database

The data for this model is typical of local government GISs. Also, sets of entities can be identified in such a diagram, as is shown in Figure 8 by the dashed rectangle used to group all entities making up the "street system."

Mapping the Conceptual Database Design to Detailed Logical/Physical Specifications

The example E-R diagram shown in Figure 8 will be used to verify with the expected users the data content of the GIS and, by additional reference to the GIS needs analysis, the required spatial operations. Once verified by the users, the E-R representation can be mapped into a detailed database design as follows:

1) Each entity and its attributes map into:
2) (a) One or more relational tables with appropriate primary and secondary keys (this assumes the desired level of normalization has been obtained); (b) The corresponding spatial entity for the "regular" entity. As most commercial GISs rely on fixed structures for the representation of geometric coordinates and topology, this step is simply reduced to ensuring that each corresponding spatial entity can be handled by the selected GIS package;

3) Each relationship into: (a) Regular relationships (diamond) executed by the relational database system's normal query structure. Again, appropriate keys and normalization are required for this mapping. (b) Spatial relationships implemented through spatial operations in the GIS. The functionality of each spatial relationship needs to be described, and if not a standard operation of the selected GIS, specifications for the indicated operation need to be written.

9

Descriptive and Geographical Data

Spatial Data Infrastructure (SDI) is an initiative intended to create an environment in which all stakeholders can co-operate with each other and interact with technology, to better achieve their objectives at different political/administrative levels. SDIs have become very important in determining the way in which spatial data are used throughout an organisation, a nation, different regions and the world. In principle, they allow the sharing of data, which is extremely useful, as it enables users to save resources, time and effort when trying to acquire new data-sets by avoiding duplication of expenses associated with generation and maintenance of data and their integration with other data-sets. By reducing duplication and facilitating integration and development of new and innovative business applications, SDIs can produce significant human and resource savings and returns.

Once formed, an SDI has often been assumed to remain a static entity. This has limited the understanding of the nature of SDIs, optimisation of their potential and the capacity for their evolution. Current perceptions and descriptions of SDI fail to convey SDI's dynamicism and complexity. Better means to describe SDI's multi-dimensional capacity as an inter-and intrajurisdictional spatial information framework are required.

The aim of this paper is to better understand and describe the nature of SDIs' currently variable identification. Researchers and various national government agencies have attempted to capture the nature of SDI in definitions produced in various contexts. This paper briefly reviews these and other definitions and argues that while they provide a useful base for the understanding of SDI, individually on their own they are inadequate for SDI development in the future.

Understanding the Nature of SDIs

While there is increasing interest being given to SDI and recognition that SDI promotes economic development and environmental management, the concept and justification of the infrastructure are still unclear. SDI remains very much an innovation even among practitioners and there are still doubts regarding the nature and identities of SDI [Barr 1998, Rajabifard, *et al.* 2000]. This is emphasised by the generally limited understanding of the innovative concept of SDI even among key players in the spatial information industry [Barr 1998, 1999, Coleman and McLaughlin 1998] all of whom are trying to understand the roles they play, past, present and future. At present, many are feeling their way as they try to position themselves in response to the continual advent of SDI.

SDI is understood differently by stakeholders from different disciplines. It is commonly recognised that an SDI can include core components of policy, fundamental datasets, technical standards, access networks and people, and adopt different design and implementation processes. In this regard, researchers and various national government agencies have attempted to capture the nature of SDI in definitions produced in various contexts. Whilst these existing definitions provide a useful base for the understanding of different aspects of SDI, or SDI at a snapshot in time, the variety of descriptions have resulted in a fragmentation of the identities and nature of SDI, derived for the varied purposes of promotion, funding and support. Lack of a more holistic representation and understanding of SDI has limited the ability to describe its evolution in response to the technical and user environment.

Existing definitions have been slow to incorporate the concept of an integrated, multi-leveled SDI. Recent research indicates that SDI is multi-leveled in nature, formed from a hierarchy of inter-connected SDIs at corporate, local, state/

provincial, national, regional (multi-national) and global levels [Chan and Williamson 1999b, Rajabifard *et al.* 1999, 2000]. SDI development at a state level also suggests that an SDI is a dynamic entity; its identity and functionality change and become more complex over time [Chan and Williamson 1999b]. Failing to acknowledge these characteristics of SDI, the multidimensionality, and dynamic mechanistic and functional roles of the SDI, have rendered many descriptions of SDI inadequate to describe the complexity and the dynamics of SDI as it develops, and thus ultimately constrain SDI achieving developmental potential in the future.

In recent years, researchers have applied the theories of *innovation diffusion* to the study of GIS planning and implementation [Onsrud and Pinto 1991, Masser 1993, Masser and Onsrud 1993, Campbell 1996, Masser and Campbell 1996, Chan 1998]. Rogers [1983] defined diffusion as a process by which an innovation is communicated through certain channels over time among the members of a social system. In 1999, Chan and Williamson [2000] applied the generic principles derived from the study of diffusion of GIS in a complex organisation to the development of SDIs. Based on the diffusion paradigm, understanding of the nature of an innovation is crucial in the success of the progressive uptake and utilisation of the innovation by members of a community [Rogers 1995]. From an engineering point of view, successful designing, building and managing an innovative product requires a sufficient understanding of the nature of the product. This is to help to establish the functions of the product and to determine the engineering characteristics of the product that best meet the customer requirements [Cross 1995].

To better understand the multidimensional nature of SDI, a system of classification is needed to organise the many definitions and various aspects of the nature of SDI. One such system has been developed to organise the definitions for GIS into four different perspectives. It treats GIS as an innovation that is progressing through a process of diffusion in an organisation [Chan and Williamson 1999a]. As GIS technology has had a significant influence on the need for SDI and the diffusion SDI is undergoing in different communities the system of definition classification for GIS should also be applicable to SDI. SDI is an innovation that is underpinned by many GIS concepts and technologies, as well as the phenomenon of the Internet and related telecommunications and network technology.

The Four Perspectives of GIS Diffusion

The definition classification system groups the definitions of GIS into four perspectives: *identificational, technological, organisational* and *productional*. The perspectives are not exclusive, rather serve descriptive purposes for different aspects of understanding GIS and at different stages of development. The *identificational* perspective describes the unique features of GIS that distinguish GIS from other types of information systems. The *technological* perspective describes GIS as a certain form of technology (database, application, or toolbox) that provides specific functional capabilities (map, database, and spatial analysis). The *organisational* perspective describes GIS in terms of its generic elements, or building blocks, which specifically include the organisational and/or institutional implementation environment. The *productional* perspective portrays GIS as the means in the production process undertaken by an organisation to generate the products and services expected by its clients.

Chan and Williamson [1999a] observe that the first three perspectives are useful at different stages of diffusion of GIS when its final purposes, functionality and composition are clearly specified—the *focused* scenario of diffusion. Under a very dynamic situation, as in the case of the long term development of a corporate GIS, when these details are not clearly defined—the *dispersed* scenario—the above three perspectives are not applicable. In this regard, an alternative composite view of GIS, namely, the *productional* perspective, was developed [Chan and Williamson 1999a].

The *productional* perspective applies to GIS in the *dispersed* scenario of diffusion. It takes a high level view of GIS in terms of the environment in which it functions and evolves. The environment is viewed as a mechanism of production. For GIS the environment generally refers to an organisation. . A module of infrastructure GIS represents the capability of one of a group of GIS suppliers in an organisation, while a business process GIS represents the GIS capability of GIS users. Inherent in the *productional* perspective is an active link or working relationship between the two modules of GIS that ensures the successful development of GIS capabilities by both GIS users and suppliers.

The diffusion of GIS in an organisation, also called corporate GIS in the literature, is viewed as the collective outcome of the individual but inter-related processes of diffusion of the GIS modules. In such a process, each GIS module can assume a different identity or even multiple identities at different stages. The outcome of diffusion of one module may have a significant impact on that of one or more other GIS modules by virtue of the links (or working relationships) established. As a result, the *productional* perspective of GIS provides a means for the mapping of the progressive development of the GIS modules and the associated links over time. It provides a better understanding of not only the nature of GIS but also the dynamics of GIS diffusion in an organisation.

The Four Perspectives for SDI

The technology of GIS has recently been incorporated in mainstream database management systems to manage spatial data along with other traditional alphanumeric data [Ower and Barrs 1999]. As GIS becomes a more mainstream corporate technology it underpins the management of spatial data in more organisations and governments. These strive to develop an integrated corporate GIS to maximise the benefit of their spatial data assets.

In the context of an SDI hierarchy, Chan and Williamson [1999b] argue that an integrated corporate GIS is in effect a corporate SDI from which SDI at different political and administrative levels can draw data. Similar to the role of SDIs, common problems addressed by a corporate GIS include elimination of duplication, acceleration of development and promotion of data sharing [Levisohn 1997]. Based on the similarities, this paper groups and discusses a range of definitions of SDI, based on the four-perspective system of classification described for GIS. The aforementioned four perspectives of classification, as applied to SDIs, are outlined in the following sub-sections.

Identificational Perspective

An *identificational* perspective describes the uniqueness of SDI. This perspective is important in justifying investment in SDI as distinct from other information infrastructure initiatives. In this regard, researchers and practitioners choose to focus on explaining the uniqueness of spatial information rather than SDI itself, as illustrated in the SDI strategies of the European Commission and the State Government of Victoria: The European Geographic Information Infrastructure (EGII) is *the European* policy framework creating the necessary conditions for achieving the objectives set out below. It thus encompasses all policies, regulations, incentives and structures set up by the EU Institutions and the Member States in this pursuit [European Commission 1995].

A spatial data infrastructure is conceptualised as *a comprehensive geospatial information resource—the infrastructure*, the value and capability of which are driven into *Victoria's information systems and processes—the benefit*, through the strategic elements of *custody, metadata, access infrastructure, pricing, spatial accuracy* and *awareness* [Victorian Geospatial Information Stategy 2000-2003 of the State Government of Victoria, Australia-Land Victoria 1999].

While useful in introducing the basic concept of SDI to the layperson, this approach runs the risk of selling short the SDI initiative to decision-makers.

Technological Perspective

A *technological* perspective describes the form and function of SDI. It provides a more tangible image of SDI in an attempt to facilitate its acceptance. A good example is the definition by McKee [1996] where the form (bold) and function (underlined) have been highlighted:

A global spatial data infrastructure is like a wheel with technology as its hub, each spoke a different country. Each country has SDI components or levels dealing with legacy data, culture, academic resources, professional organisations, governmental agencies, and legal and regulatory structures (for *land tenure, privacy, intellectual property, environment, census, etc.).*

Another example is the portrayal of an ideal SDI as a *hierarchy of spatial datasets* that users at different levels can access to meet their needs. The Canadian Geospatial Data Infrastructure (CGDI) is also defined using a technological perspective as the *technology, standards, access systems and protocols necessary to harmonize all of Canada's geospatial data bases, and make them available on the internet* [CGDI 2000]. Three key elements that have gained a high profile in the literature in recent years are the framework data, standards and the delivery mechanism of SDI. Though these elements are only part of the whole SDI, they are often portrayed as one of the several simplified identities of SDI depending on the strategies of the SDI managers. For example, regarding framework data, The framework forms the data backbone of the NSDI and is designed to facilitate production and use of geographic data, to reduce operating costs, *and to improve service and decision making.*

The delivery mechanism of SDI is primarily made up of a component that allows searching of spatial data and another that allows browsing and down-loading spatial data over a network, often the Internet. The former is often called a spatial data directory. In the case of Australia, The Australian Spatial Data Directory (ASDD) is an essential component of the Australian Spatial Data Infrastructure (ASDI) and provides these [sic] search interfaces to geospatial dataset descriptions (metadata) from all jurisdictions *throughout Australia* [ANZLIC Metadata Working Group 1999]. The latter is often called a clearinghouse. According to FGDC [1999], The clearinghouse is a decentralized system of servers located on the Internet which ... provide(s) access to digital spatial data through metadata.

Organisational Perspective

An *organisational* perspective describes SDI in terms of its building blocks [Rajabifard *et al.* 1999, 2000] and in particular the organisational/institutional setting. It is meant to allow

SDI practitioners to better plan and coordinate the development of the very complex innovation of SDI.

The components of a spatial data infrastructure should include *sources of spatial data,* databases and metadata, data networks, technology (dealing with data collection, management and representation), institutional arrangements, policies and standards and *end-users* [McLaughlin and Nichols 1992].

Other governments such as the Netherlands [Masser 1998], the State of Queensland [Department of Natural Resources 1999] and international organisations such as the Global Spatial Data Infrastructure [GSDI 1999], opt for a composite approach —an amalgamation of the *technological* and *organisational* perspectives. In this approach, the form of SDI is described in terms of the building blocks such as those put forward by McLaughlin and Nichols [1992] while the function of the SDI is described as serving the needs of the stakeholders concerned.

Productional Perspective

Like that of GIS, the diffusion of SDI also takes place in a *dispersed* scenario in which the final purposes, functionality and composition of the SDI are only vaguely defined as illustrated in the strategies of SDI development of different countries [European Commission 1995, ANZLIC 1996, FGDC 1999]. The experience with GIS suggests that from the similarities of characteristics of an SDI to a GIS, taking a *productional* perspective for SDI may provide insights into the characteristics of SDI in the context of its evolving environment, resulting in more holistic strategies to manage its diffusion and development within that environment. Chan and Williamson [1999b] propose that such an approach is justified given the characteristics of an SDI are similar to those of a GIS. Based on the experience of development of GIS in the State of Victoria in Australia and Australia as a whole, Chan and Williamson [1999b] cite similarities between SDIs and GIS to include:

1. Like a module of *infrastructure GIS,* an SDI does not develop in isolation but in conjunction with the business activities it supports;
2. It is a multi-levelled entity functionally and administratively;

3. It is dynamic in nature and even well conceived centralised planning will not guarantee the development of an SDI that meets the needs of all users.

As section 3 points out, the two roles of a module of GIS depict the interdependency of the suppliers and users of GIS in an organisation. Likewise, the *productional* perspective of SDI should describe an SDI in terms of a dual-roled module in an environment where business activities or production processes, underpinned by GIS, are undertaken. Chan and Williamson [1999b] propose that a corporate GIS, also constitute a corporate SDI which is a building block of an SDI hierarchy that spans different political/administrative levels throughout the world. Based on this concept the environment for the corporate SDI is identified as the organisation to which it belongs. Each organisation, and the corporate SDI it establishes, can play the role of a supplier or user of spatial data needed to conduct business in a jurisdiction. This concept help to describe the spatial information industry suggested above-the environment for SDIs at other levels in the SDI hierarchy. In the model the building blocks are individual corporate SDIs which use and/or supply spatial data and technology, and interact progressively with one another as members of the spatial information industry, in the production processes of a jurisdiction to fulfil its range of business needs-social, economic and environmental.

The interaction involves spatial data and technology users adding value to the original raw data provided by spatial data suppliers, and then on-selling the value-added data to other users. This continual value-adding process ends at an ultimate user, often a member of the public, who uses a spatial data product to make decisions. As a result, the interacting stakeholders groups can be visualised as a network of value-adding chains of suppliers and users of spatial data/technology, or alternatively, the different dimensions of the spatial information industry. This industry, in turn, represents the environment in which the SDI functions and evolves.

It takes into consideration the interests of both spatial data users and suppliers and the way they interact to deliver the products and services to meet the jurisdiction's business needs.

As the reach of the industry is worldwide and recognises no political or administrative boundary, it represents an environment that is generic enough to account for the hierarchy of SDI observed.

Conclusions

This paper points out that SDI remains an innovation among many practitioners. There is a need to better clarify the nature of SDI to facilitate its development and progressive uptake and utilisation among members of a community (diffusion). A number of the more current definitions of SDI are reviewed within a classification system of four perspectives of SDI, namely, *identificational, technological, organisational* and *productional*. The definitions fall within the first three perspectives with the organisational perspective being the most popular approach adopted by government, regional and global SDI developing agencies. Based on the research into diffusion of corporate GIS, it is proposed that SDI development and diffusion takes place in a *dispersed* scenario in which the final purposes, functionalities and composition of the SDI change dynamically and can only be specified vaguely. Under this condition, it is the fourth perspective, the *productional* perspective, not the first three perspectives of SDI, that is potentially most useful in facilitating SDI development and diffusion.

Statistical Inference

Statistical inference is the process of making conclusions using data that is subject to random variation, for example, observational errors or sampling variation. More substantially, the terms statistical inference, statistical induction and inferential statistics are used to describe systems of procedures that can be used to draw conclusions from datasets arising from systems affected by random variation. Initial requirements of such a system of procedures for inference and induction are that the system should produce reasonable answers when applied to well-defined situations and that it should be general enough to be applied across a range of situations.

The outcome of statistical inference may be an answer to the question "what should be done next?", where this might be a decision about making further experiments or surveys, or about drawing a conclusion before implementing some organizational or governmental policy.

Introduction

Scope

For the most part, statistical inference makes propositions about populations, using data drawn from the population of interest via some form of random sampling. More generally, data about a random process is obtained from its observed behaviour during a finite period of time. Given a parameter or hypothesis about which one wishes to make inference, statistical inference most often uses:

- a statistical model of the random process that is supposed to generate the data, and
- a particular realization of the random process; i.e., a set of data.

The conclusion of a statistical inference is a statistical proposition. Some common forms of statistical proposition are:

- an estimate; i.e., a particular value that best approximates some parameter of interest,
- a confidence interval (or set estimate); i.e., an interval constructed from the data in such a way that, under repeated sampling of datasets, such intervals would contain the true parameter value with the probability at the stated confidence level,
- a credible interval; i.e., a set of values containing, for example, 95% of posterior belief,
- rejection of an hypothesis
- clustering or classification of data points into groups

Comparison to Descriptive Statistics

Statistical inference is generally distinguished from descriptive statistics. In simple terms, descriptive statistics can be thought of as being just a straightforward presentation of facts, in which modelling decisions made by a data analyst have had minimal influence. A complete statistical analysis will nearly always include both descriptive statistics and statistical inference, and will often progress in a series of steps where the emphasis moves gradually from description to inference.

Models/Assumptions

Any statistical inference requires some assumptions. A statistical model is a set of assumptions concerning the generation of the observed data and similar data. Descriptions of statistical models usually emphasize the role of population quantities of interest, about which we wish to draw inference.

Degree of Models/Assumptions

Statisticians distinguish between three levels of modelling assumptions;

- Fully parametric: The probability distributions describing the data-generation process are assumed to be fully described by a family of probability distributions involving only a finite number of unknown parameters. For example, one may assume that the distribution of population values is truly Normal, with unknown mean and variance, and that datasets are generated by 'simple' random sampling. The family of generalized linear models is a widely-used and flexible class of parametric models.
- Non-parametric: The assumptions made about the process generating the data are much less than in parametric statistics and may be minimal.. For example, every continuous probability distribution has a median, which may be estimated using the sample median or theHodges-Lehmann-Sen estimator, which has good properties when the data arise from simple random sampling.
- Semi-parametric: This term typically implies assumptions 'between' fully and non-parametric approaches. For example, one may assume that a

population distribution have a finite mean. Furthermore, one may assume that the mean response level in the population depends in a truly linear manner on some covariate (a parametric assumption) but not make any parametric assumption describing the variance around that mean (i.e., about the presence or possible form of any heteroscedasticity). More generally, semi-parametric models can often be separated into 'structural' and 'random variation' components. One component is treated parametrically and the other non-parametrically. The well-known Cox model is a set of semi-parametric assumptions.

Importance of Valid Models/Assumptions

Whatever level of assumption is made, correctly-calibrated inference in general requires these assumptions to be correct; i.e., that the data-generating mechanisms really has been correctly specified.

Incorrect assumptions of 'simple' random sampling can invalidate statistical inference . More complex semi-and fully-parametric assumptions are also cause for concern. For example, incorrectly assuming the Cox model can in some cases lead to faulty conclusions . Incorrect assumptions of Normality in the population also invalidates some forms of regression-based inference . The use of any parametric model is viewed skeptically by most experts in sampling human populations: "most sampling statisticians, when they deal with confidence intervals at all, limit themselves to statements about [estimators] based on very large samples, where the central limit theorem ensures that these [estimators] will have distributions that are nearly normal." Here, the central limit theorem states that the distribution of the sample mean "for very large samples" is approximately normally distributed, if the distribution is not heavy tailed.

Approximate Distributions

Given the difficulty in specifying exact distributions of sample statistics, many methods have been developed for approximating these.

With finite samples, approximation results measure how close a limiting distribution approaches the statistic's sample distribution: For example, with 10,000 independent samples the normal distribution approximates (to two digits of accuracy) the distribution of the sample mean for many population distributions, by the Berry–Esseen theorem.

Yet for many practical purposes, the normal approximation provides a good approximation to the sample-mean's distribution when there are 10 (or more) independent samples, according to simulation studies, and statisticians' experience.

Following Kolmogorov's work in the 1950s, advanced statistics uses approximation theory andfunctional analysis to quantify the error of approximation: In this approach, the metric geometry of probability distributions is studied; this approach quantifies approximation error with, for example, the Kullback–Leibler distance, Bregman divergence, and the Hellinger distance.

With infinite samples, limiting results like the central limit theorem describe the sample statistic's limiting distribution, if one exists. Limiting results are not statements about finite samples, and indeed are logically irrelevant to finite samples. However, the asymptotic theory of limiting distributions is often invoked for work in estimation and testing.

For example, limiting results are often invoked to justify thegeneralized method of moments and the use of generalized estimating equations, which are popular in econometrics and biostatistics. The magnitude of the difference between the limiting distribution and the true distribution (formally, the 'error' of the approximation) can be assessed using simulation:. The use of limiting results in this way works well in many applications, especially with low-dimensionalmodels with log-concave likelihoods (such as with one-parameter exponential families).

Randomization-based Models

For a given dataset that was produced by a randomization design, the randomization distribution of a statistic (under the null-hypothesis) is defined by evauating the test statistic for all of the plans that could have been generated by the randomiation design.

In frequentist inference, randomization allows inferences to be based on the randomization distribution rather than a subjective model, and this is important especially in survey sampling and design of experiments.. Statistical inference from randomized studies is also more straightforward than many other situations. In Bayesian inference, randomization is also of importance: In survey sampling – sampling without replacementensures the exchangeability of the sample with the population; in randomized experiments, randomization warrants a missing at randomassumption for covariate information.

Objective randomization allows properly inductive procedures. Many statisticians prefer randomization-based analysis of data that was generated by well-defined randomization procedures. (However, it is true that in fields of science with developed theoretical knowledge and experimental control, randomized experiments may increase the costs of experimentation without improving the quality of inferences.)

Similarly, results from randomized experiments are recommended by leading statistical authorities as allowing inferences with greater reliability than do observational studies of the same phenomena. However, a good observational study may be better than a bad randomized experiment.

The statistical analysis of a randomized experiment may be based on the randomization scheme stated in the experimental protocol and does not need a subjective model.

However, not all hypotheses can be tested by randomized experiments or random samples, which often require a large budget, a lot of expertise and time, and may have ethical problems.

Model-based Analysis of Randomized Experiments

It is standard practice to refer to a statistical model, often a normal linear model, when analyzing data from randomized experiments. However, the randomization scheme guides the choice of a statistical model. It is not possible to choose an appropriate model without knowing the randomization scheme. Seriously misleading results can be obtained analyzing data from randomized experiments while ignoring the experimental protocol; common mistakes include forgetting the blocking used in an experiment and confusing repeated measurements on the same experimental unit with independent replicates of the treatment applied to different experimental units.

Modes of Inference

Different schools of statistical inference have become established. These schools (or 'paradigms') are not mutually-exclusive, and methods which work well under one paradigm often have attractive interpretations under other paradigms. The two main paradigms in use arefrequentist and Bayesian inference, which are both summarized below.

Frequentist Inference

This paradigm calibrates the production of propositions by considering (notional) repeated sampling of datasets similar to the one at hand. By considering its characteristics under repeated sample, the frequentist properties of any statistical inference procedure can be described-although in practice this quantification may be challenging.

Examples of Frequentist Inference

- P-value
- Confidence interval.

Frequentist Inference, Objectivity, and Decision Theory

Frequentist inference calibrates procedures, such as tests of hypothesis and constructions of confidence intervals, in terms of frequency probability; that is, in terms of repeated sampling from a population. (In contrast, Bayesian inference calibrates procedures with regard to epistemological uncertainty, described as a probability measure)

The frequentist calibration of procedures can be done without regard to utility functions. However, some elements of frequentist statistics, such as statistical decision theory, do incorporate utility functions. In particular, frequentist developments of optimal inference (such as minimum-variance unbiased estimators, or uniformly most powerful testing) make use of loss functions, which play the role of (negative) utility functions. Loss functions must be explicitly stated for statistical theorists to prove that a statistical procedure has an optimality property. For example, median-unbiased estimators are optimal under absolute value loss functions, and least squares estimators are optimal under squared error loss functions.

While statisticians using frequentist inference must choose for themselves the parameters of interest, and the estimators/ test statistic to be used, the absence of obviously-explicit utilities and prior distributions has helped frequentist procedures to become widely-viewed as 'objective'.

Bayesian Inference

The Bayesian calculus describes degrees of belief using the 'language' of probability; beliefs are positive, integrate to one, and obey probability axioms. Bayesian inference uses the available posterior beliefs as the basis for making statistical propositions. There are several different justifications for using the Bayesian approach.

Examples of Bayesian Inference

- Credible intervals for interval estimation
- Bayes factors for model comparison.

Bayesian Inference, Subjectivity and Decision Theory

Many informal Bayesian inferences are based on "intuitively reasonable" summaries of the posterior. For example, the posterior mean, median and mode, highest posterior density intervals, and Bayes Factors can all be motivated in this way. While a user's utility function need not be stated for this sort of inference, these summaries do all depend (to some extent) on stated prior beliefs, and are generally viewed as subjective conclusions. (Methods of prior construction which do not require external input have been proposed but not yet fully developed.)

Formally, Bayesian inference is calibrated with reference to an explicitly stated utility, or loss function; the 'Bayes rule' is the one which maximizes expected utility, averaged over the posterior uncertainty. Formal Bayesian inference therefore automatically provides optimal decisions in a decision theoretic sense.

Given assumptions, data and utility, Bayesian inference can be made for essentially any problem, although not every statistical inference need have a Bayesian interpretation. Analyses which are not formally Bayesian can be (logically)incoherent; a feature of Bayesian procedures which use proper priors (i.e., those integrable to one) is that they are guaranteed to becoherent. Some advocates of Bayesian inference assert that inference *must* take place in this decision-theoretic framework, and thatBayesian inference should not conclude with the evaluation and summarization of posterior beliefs.

Other Modes of Inference (Besides Frequentist and Bayesian)

Information and Computational Complexity

Other forms of statistical inference have been developed from ideas in information theory and the theory of Kolmogorov complexity.. For example, the minimum description length (MDL) principle selects statistical models that maximally compress the data; inference proceeds without assuming counterfactual or non-falsifiable 'data-generating mechanisms' or probability models for the data, as might be done in frequentist or Bayesian approaches.

However, if a 'data generating mechanism' does exist in reality, then according to Shannon's source coding theorem it provides the MDL description of the data, on average and asymptotically. In minimizing description length (or descriptive complexity), MDL estimation is similar to maximum likelihood estimation and maximum a posteriori estimation (using maximum-entropy Bayesian priors).

However, MDL avoids assuming that the underlying probability model is known; the MDL principle can also be applied without assumptions that e.g. the data arose from independent sampling. The MDL principle has been applied in communication-coding theory in information theory, in linear regression, and in time-series analysis (particularly for chosing the degrees of the polynomials in Autoregressive moving average (ARMA) models).

Information-theoretic statistical inference has been popular in data mining, which has become a common approach for very large observational and heterogeneous datasets made possible by the computer revolution and internet.

The evaluation of statistical inferential procedures often uses techniques or criteria from computational complexity theory or numerical analysis.

Fiducial Inference

Fiducial inference was an approach to statistical inference based on fiducial probability, also known as a "fiducial distribution". In subsequent work, this approach has been recognized as being ill-defined, extremely limited in applicability, and even fallacious.

Structural Inference

Developing ideas of Fisher and of Pitman from 1938-1939, George A. Barnard developed "structural inference" or "pivotal inference", an approach using invariant probabilities on group families. Barnard reformulated the arguments behind fiducial inference on a restricted class of models on which "fiducial" procedures would be well-defined and useful.

1. Statistical assumptions
2. Statistical decision theory
3. Estimation theory
4. Statistical hypothesis testing
5. Revising opinions in statistics
6. Design of experiments, the analysis of variance, and regression
7. Survey sampling
8. Summarizing statistical data.

Descriptive statistics are used to describe the main features of a collection of data quantitatively. Descriptive statistics are distinguished from inferential statistics (or inductive statistics), in that descriptive statistics aim to summarize a data set quantitatively without employing a probabilistic formulation, rather than being used to support inferential statements about the population that the data are thought to represent.

Even when a data analysis draws its main conclusions using inductive statistical analysis, descriptive statistics are generally presented along with more formal analyses.

For example in a paper reporting on a study involving human subjects, there typically appears a table giving the overall sample size, sample sizes in important subgroups (e.g., for each treatment or exposure group), and demographic or clinical characteristics such as the average age, the proportion of subjects of each sex, and the proportion of subjects with relatedcomorbidities.

In research involving comparisons between groups, a major emphasis is often placed on the significance level for the hypothesis that the groups being compared differ to a greater degree than would be expected by chance.

This significance level is often represented as a p-value, or sometimes as the standard score of a test statistic. In contrast, an effect size is a descriptive statistic that conveys the estimated magnitude and direction of the difference between groups, without regard to whether the difference is statistically significant. Reporting significance levels without effect sizes is often criticized, since for large sample sizes even small effects of little practical importance can be highly statistically significant.

10

Generalized Linear Models for Continuous Data

In statistics, the generalized linear model (GLM) is a flexible generalization of ordinary least squares regression. The GLM generalizes linear regression by allowing the linear model to be related to the response variable via a link function and by allowing the magnitude of the variance of each measurement to be a function of its predicted value.

Generalized linear models were formulated by John Nelder and Robert Wedderburn as a way of unifying various other statistical models, including linear regression, logistic regression and Poisson regression. They proposed an iteratively reweighted least squares method for maximum likelihood estimation of the model parameters. Maximum-likelihood estimation remains popular and is the default method on many statistical computing packages. Other approaches, including Bayesian approaches and least squares fits to variance stabilized responses, have been developed.

Basic Ideas

The Generalized Linear Model (GLZ) is a generalization of the general linear model. In its simplest form, a linear model specifies the (linear) relationship between a dependent (or response) variable *Y*, and a set of predictor variables, the *X*'s, so that

$$Y = b_0 + b_1X_1 + b_2X_2 + ... + b_kX_k$$

In this equation b_0 is the regression coefficient for the intercept and the b_i values are the regression coefficients (for variables 1 through *k*) computed from the data.

So for example, we could estimate (i.e., predict) a person's weight as a function of the person's height and gender. You could use linear regression to estimate the respective regression coefficients from a sample of data, measuring height, weight, and observing the subjects' gender. For many data analysis problems, estimates of the linear relationships between variables are adequate to describe the observed data, and to make reasonable predictions for new observations

However, there are many relationships that cannot adequately be summarized by a simple linear equation, for two major reasons:

Distribution of dependent variable. First, the dependent variable of interest may have a non-continuous distribution, and thus, the predicted values should also follow the respective distribution; any other predicted values are not logically possible. For example, a researcher may be interested in predicting one of three possible discrete outcomes (e.g., a consumer's choice of one of three alternative products). In that case, the dependent variable can only take on 3 distinct values, and the distribution of the dependent variable is said to be *multinomial.* Or suppose you are trying to predict people's family planning choices, specifically, how many children families will have, as a function of income and various other socioeconomic indicators. The dependent variable-number of children-is discrete (i.e., a family may have 1, 2, or 3 children and so on, but cannot have 2.4 children), and most likely the distribution of that variable is highly skewed (i.e., most families have 1, 2, or 3 children, fewer will have 4 or 5, very few will have 6 or 7, and so on). In this case it would be reasonable to assume that the dependent variable follows a Poisson distribution.

Link function. A second reason why the linear (multiple regression) model might be inadequate to describe a particular relationship is that the effect of the predictors on the dependent variable may not be linear in nature. For example, the relationship between a person's age and various indicators of health is most likely not linear in nature: During early adulthood, the (average) health status of people who are 30 years old as compared to the (average) health status of people who are 40 years old is not markedly different. However, the difference in health status of 60 year old people and 70 year old people is probably greater. Thus, the relationship between age and health status is likely non-linear in nature. Probably some kind of a power function would be adequate to describe the relationship between a person's age and health, so that each increment in years of age at older ages will have greater impact on health status, as compared to each increment in years of age during early adulthood. Put in other words, the *link* between age and health status is best described as non-linear, or as a power relationship in this particular example.

The generalized linear model can be used to predict responses both for dependent variables with discrete distributions and for dependent variables which are nonlinearly related to the predictors.

Computational Approach

To summarize the *basic ideas*, the generalized linear model differs from the general linear model (of which, for example, multiple regression is a special case) in two major respects: First, the distribution of the dependent or response variable can be (explicitly) non-normal, and does not have to be continuous, i.e., it can be binomial, multinomial, or ordinal multinomial (i.e., contain information on ranks only); second, the dependent variable values are predicted from a linear combination of predictor variables, which are "connected" to the dependent variable via a link function. The general linear model for a single dependent variable can be considered a special case of the generalized linear model: In the general linear model the dependent variable values are expected to follow the normal distribution, and the link function is a simple identity function (i.e., the linear combination of values for the predictor variables is not transformed).

To illustrate, in the general linear model a response variable *Y* is linearly associated with values on the *X* variables by

$$Y = b_0 + b_1X_1 + b_2X_2 + ... + b_kX_k + e$$

(where *e* stands for the error variability that cannot be accounted for by the predictors; note that the expected value of *e* is assumed to be 0), while the relationship in the generalized linear model is assumed to be

$$Y = g\ (b_0 + b_1X_1 + b_2X_2 + \ldots + b_kX_k) + e$$

where *e* is the error, and *g(...)* is a function. Formally, the inverse function of *g(...)*, say *f(...)*, is called the link function; so that:

$$f(mu_y) = b_0 + b_1X_1 + b_2X_2 + \ldots + b_kX_k$$

where mu_y stands for the expected value of *y*.

Link functions and distributions. Various link functions can be chosen, depending on the assumed distribution of the *y* variable values:

Estimation in the generalized linear model. The values of the parameters (b_0 through b_k and the scale parameter) in the generalized linear model are obtained by maximum likelihood (ML) estimation, which requires iterative computational procedures. There are many iterative methods for ML estimation in the generalized linear model, of which the Newton-Raphson and Fisher-Scoring methods are among the most efficient and widely used. The Fisher-scoring (or iterative re-weighted least squares) method in particular provides a unified algorithm for all generalized linear models, as well as providing the expected variance-covariance matrix of parameter estimates as a byproduct of its computations.

Statistical significance testing. Tests for the significance of the effects in the model can be performed via the Wald statistic, the likelihood ratio (LR), or score statistic. Detailed descriptions of these tests can be found in McCullagh and Nelder (1989). The Wald statistic, which is computed as the generalized inner product of the parameter estimates with the respective variance-covariance matrix, is an easily computed, efficient statistic for testing the significance of effects. The score statistic is obtained from the generalized inner product of the score vector with the Hessian matrix (the matrix of the second-order partial derivatives of the maximum likelihood parameter estimates). The likelihood ratio (LR) test requires the greatest computational effort (another iterative estimation procedure) and is thus not as fast as the first two methods; however, the LR test provides the most asymptotically efficient test known.

Diagnostics in the generalized linear model. The two basic types of residuals are the so-called Pearson residuals and deviance residuals. Pearson residuals are based on the difference between observed responses and the predicted values; deviance residuals are based on the contribution of the observed responses to the log-likelihood statistic. In addition, leverage scores, studentized residuals, generalized Cook's D, and other observational statistics (statistics based on individual observations) can be computed. For a description and discussion of these statistics.

Model Building

In addition to fitting the whole model for the specified type of analysis, different methods for automatic model building can be employed in analyses using the generalized linear model. Specifically, forward entry, backward removal, forward stepwise, and backward stepwise procedures can be performed, as well as best-subset search procedures. In forward methods of selection of effects to include in the model (i.e., forward entry and forward stepwise methods), score statistics are compared to select new (significant) effects. The Wald statistic can be used for backward removal methods (i.e., backward removal and backward stepwise, when effects are selected for removal from the model).

The best subsets search method can be based on three different test statistics: the score statistic, the model likelihood, and the AIC. Note that, since the score statistic does not require iterative computations, best subset selection based on the score statistic is computationally fastest.

Interpretation of Results and Diagnostics

Simple estimation and test statistics may not be sufficient for adequate interpretation of the effects in an analysis. Especially for higher order (e.g., interaction) effects, inspection of the observed and predicted means can be invaluable for understanding the nature of an effect. Plots of these means (with error bars) can be useful for quickly grasping the role of the effects in the model.

Inspection of the distributions of variables is critically important when using the generalized linear model. Histograms and probability plots for variables, and scatterplots showing the relationships between observed values, predicted values, and residuals (e.g., Pearson residuals, deviance residuals, studentized residuals, differential *Chi-square* statistics, differential deviance statistics, and generalized Cook's D) provide invaluable model-checking tools.

Multiple Regression

The general purpose of multiple regression (the term was first used by Pearson, 1908) is to learn more about the relationship between several independent or predictor variables and a dependent or criterion variable. For example, a real estate agent might record for each listing the size of the house (in square feet), the number of bedrooms, the average income in the respective neighborhood according to census data, and a subjective rating of appeal of the house. Once this information has been compiled for various houses it would be interesting to see whether and how these measures relate to the price for which a house is sold. For example, you might learn that the number of bedrooms is a better predictor of the price for which a house sells in a particular neighborhood than how "pretty" the house is (subjective rating). You may also detect "outliers," that is, houses that should really sell for more, given their location and characteristics.

Personnel professionals customarily use multiple regression procedures to determine equitable compensation. You can determine a number of factors or dimensions such as "amount of responsibility" (*Resp*) or "number of people to supervise" (*No_Super*) that you believe to contribute to the value of a job. The personnel analyst then usually conducts a salary survey among comparable companies in the market, recording the salaries and respective characteristics (i.e., values on dimensions) for different positions. This information can be used in a multiple regression analysis to build a regression equation of the form:

Salary =.5*Resp +.8*No_Super

Once this so-called regression line has been determined, the analyst can now easily construct a graph of the expected (predicted) salaries and the actual salaries of job incumbents in his or her company. Thus, the analyst is able to determine which position is underpaid or overpaid, or paid equitably.

In the social and natural sciences multiple regression procedures are very widely used in research. In general, multiple regression allows the researcher to ask (and hopefully answer) the general question "what is the best predictor of...". For example, educational researchers might want to learn what are the best predictors of success in high-school. Psychologists may want to determine which personality variable best predicts social adjustment. Sociologists may want to find out which of the multiple social indicators best predict whether or not a new immigrant group will adapt and be absorbed into society.

Computational Approach

The general computational problem that needs to be solved in multiple regression analysis is to fit a straight line to a number of points.

Least Squares

In the scatterplot, we have an independent or X variable, and a dependent or Y variable. These variables may, for example, represent IQ (intelligence as measured by a test) and school achievement (grade point average; GPA), respectively. Each point in the plot represents one student, that is, the respective student's IQ and GPA. The goal of linear regression procedures is to fit a line through the points. Specifically, the program will compute a line so that the squared deviations of the observed points from that line are minimized. Thus, this general procedure is sometimes also referred to as least squares estimation.

The Regression Equation

A line in a two dimensional or two-variable space is defined by the equation $Y=a+b*X$; in full text: the Y variable can be expressed in terms of a constant (a) and a slope (b) times the

X variable. The constant is also referred to as the *intercept*, and the slope as the *regression coefficient* or *B coefficient*. For example, GPA may best be predicted as *1+.02*IQ*. Thus, knowing that a student has an *IQ* of 130 would lead us to predict that her GPA would be 3.6 (since, 1+.02*130=3.6).

In the multivariate case, when there is more than one independent variable, the regression line cannot be visualized in the two dimensional space, but can be computed just as easily. For example, if in addition to *IQ* we had additional predictors of achievement (e.g., *Motivation, Self-discipline*) we could construct a linear equation containing all those variables. In general then, multiple regression procedures will estimate a linear equation of the form:

$$Y = a + b_1 {*} X_1 + b_2 {*} X_2 + ... + b_p {*} X_p$$

Unique Prediction and Partial Correlation

Note that in this equation, the regression coefficients (or *B* coefficients) represent the *independent* contributions of each independent variable to the prediction of the dependent variable. Another way to express this fact is to say that, for example, variable X_1 is correlated with the *Y* variable, after controlling for all other independent variables. This type of correlation is also referred to as a *partial correlation* (this term was first used by Yule, 1907). Perhaps the following example will clarify this issue. You would probably find a significant negative correlation between hair length and height in the population (i.e., short people have longer hair).

At first this may seem odd; however, if we were to add the variable *Gender* into the multiple regression equation, this correlation would probably disappear. This is because women, on the average, have longer hair than men; they also are shorter on the average than men. Thus, after we remove this gender difference by entering *Gender* into the equation, the relationship between hair length and height disappears because hair length does *not* make any unique contribution to the prediction of height, above and beyond what it shares in the prediction with variable *Gender*. Put another way, after controlling for the variable *Gender*, the partial correlation between hair length and height is zero.

Predicted and Residual Scores

The regression line expresses the best prediction of the dependent variable (*Y*), given the independent variables (*X*). However, nature is rarely (if ever) perfectly predictable, and usually there is substantial variation of the observed points around the fitted regression line (as in the scatterplot shown earlier). The deviation of a particular point from the regression line (its predicted value) is called the *residual* value.

Residual Variance and R-square

R-Square, also known as the *Coefficient of determination* is a commonly used statistic to evaluate model fit. *R-square* is 1 minus the *ratio of residual variability*. When the variability of the residual values around the regression line relative to the overall variability is small, the predictions from the regression equation are good. For example, if there is no relationship between the X and Y variables, then the *ratio of the residual variability* of the Y variable to the original variance is equal to 1.0. Then R-square would be 0. If X and Y are perfectly related then there is no residual variance and the ratio of variance would be 0.0, making R-square = 1. In most cases, the ratio and *R-square* will fall somewhere between these extremes, that is, between 0.0 and 1.0. This ratio value is immediately interpretable in the following manner. If we have an *R-square* of 0.4 then we know that the variability of the Y values around the regression line is 1-0.4 times the original variance; in other words we have explained 40% of the original variability, and are left with 60% residual variability. Ideally, we would like to explain most if not all of the original variability. The *R-square* value is an indicator of how well the model fits the data (e.g., an *R-square* close to 1.0 indicates that we have accounted for almost all of the variability with the variables specified in the model).

Interpreting the Correlation Coefficient R

Customarily, the degree to which two or more predictors (independent or *X* variables) are related to the dependent (*Y*) variable is expressed in the correlation coefficient *R*, which is the square root of *R-square*. In multiple regression, *R* can assume values between 0 and 1. To interpret the direction of the relationship between variables, look at the signs (plus or minus) of the regression or *B* coefficients. If a *B* coefficient is positive, then the relationship of this variable with the dependent variable is positive (e.g., the greater the IQ the better the grade point average); if the *B* coefficient is negative then the relationship is negative (e.g., the lower the class size the better the average test scores). Of course, if the *B* coefficient is equal to 0 then there is no relationship between the variables.

Assumptions, Limitations, Practical Considerations

Assumption of Linearity

First of all, as is evident in the name multiple *linear* regression, it is assumed that the relationship between variables is linear. In practice this assumption can virtually never be confirmed; fortunately, multiple regression procedures are not greatly affected by minor deviations from this assumption. However, as a rule it is prudent to *always* look at bivariate scatterplot of the variables of interest. If curvature in the relationships is evident, you may consider either transforming the variables, or explicitly allowing for nonlinear components.

Normality Assumption

It is assumed in multiple regression that the residuals (predicted minus observed values) are distributed normally (i.e., follow the normal distribution). Again, even though most tests (specifically the *F*-test) are quite robust with regard to violations of this assumption, it is *always* a good idea, before drawing final conclusions, to review the distributions of the major variables of interest. You can produce histograms for the residuals as well as normal probability plots, in order to inspect the distribution of the residual values.

Limitations

The major conceptual limitation of all regression techniques is that you can only ascertain *relationships*, but never be sure about underlying *causal* mechanism. For example, you would find a strong positive relationship (correlation) between the damage that a fire does and the number of firemen involved in fighting the blaze. Do we conclude that the firemen cause the damage? Of course, the most likely explanation of this correlation is that the size of the fire (an external variable that we forgot to include in our study) caused the damage as well as the involvement of a certain number of firemen (i.e., the bigger the fire, the more firemen are called to fight the blaze). Even though this example is fairly obvious, in real correlation research, alternative causal explanations are often not considered.

Choice of the Number of Variables

Multiple regression is a seductive technique: "plug in" as many predictor variables as you can think of and usually at least a few of them will come out significant. This is because you are capitalizing on chance when simply including as many variables as you can think of as predictors of some other variable of interest. This problem is compounded when, in addition, the number of observations is relatively low. Intuitively, it is clear that you can hardly draw conclusions from an analysis of 100 questionnaire items based on 10 respondents. Most authors recommend that you should have at least 10 to 20 times as many observations (cases, respondents) as you have variables; otherwise the estimates of the regression line are probably very unstable and unlikely to replicate if you were to conduct the study again.

Multicollinearity and Matrix Ill-Conditioning

This is a common problem in many correlation analyses. Imagine that you have two predictors (*X* variables) of a person's height: (1) weight in pounds and (2) weight in ounces. Obviously, our two predictors are completely redundant; weight is one and the same variable, regardless of whether it is measured in pounds or ounces. Trying to decide which one of the two measures is a better predictor of height would be rather silly; however, this is exactly what you would try to do if you were to perform a multiple regression analysis with height as the dependent (*Y*) variable and the two measures of weight as the independent (X) variables. When there are very many variables involved, it is often not immediately apparent that this problem exists, and it may only manifest itself after several variables have already been entered into the regression equation. Nevertheless, when this problem occurs it means that at least one of the predictor variables is (practically) completely redundant with other predictors. There are many statistical indicators of this type of redundancy (tolerances, semi-partial *R*, etc., as well as some remedies (e.g., *Ridge regression*).

Fitting Centred Polynomial Models

The fitting of higher-order polynomials of an independent variable with a mean not equal to zero can create difficult multicollinearity problems. Specifically, the polynomials will be highly correlated due to the mean of the primary independent variable. With large numbers (e.g., Julian dates), this problem is very serious, and if proper protections are not put in place, can cause wrong results.

The Importance of Residual Analysis

Even though most assumptions of multiple regression cannot be tested explicitly, gross violations can be detected and should be dealt with appropriately. In particular outliers (i.e., extreme cases) can seriously bias the results by "pulling" or "pushing" the regression line in a particular direction, thereby leading to biased regression coefficients. Often, excluding just a single extreme case can yield a completely different set of results.

Bibliography

Allen, Kathleen M. S., Stanton W. Green: *Interpreting Space: GIS and Archaeology*, London and New York: Taylor & Francis, 1990.

Balshaw, Maria, and Liam Kennedy: *Urban Space and Representation.* Sterling, VA: Pluto Press, 2000.

Belden C.: *Landscapes of the Sacred: Geography and Narrative in American Spirituality*, New York: Paulist Press, 1988.

Berry, J.K. : *Beyond Mapping, Concepts, Algorithms and Issues in GIS*, Fort Collins, CO: GIS World Books, 1993.

Bolstad, P. : *GIS Fundamentals, A first text on Geographic Information Systems, Second Edition*, White Bear Lake, MN, Eider Press, 2005.

Burrough, P.A. and McDonnell, R.A. : *Principles of Geographical Information Systems*, Oxford University Press, Oxford, 1998.

Carrillo, J., Lung, Y., and van Tulder, R. : *Cars, Carriers of Regionalism*, Palgrave Macmillan, New York, 2004.

Cronon, William: *Changes in the Land: Indians, Colonists, and the Ecology of New England.* New York: Hill and Wang, 1983.

Davis, Mike: *Ecology of Fear: Los Angeles and the Imagination of Disaster.* New York: Vintage, 1999.

Dibbern, Jens : *The Sourcing of Application Software Services, Empirical Evidence of Cultural, Industry and Functional Differences,* Physica Verlag Heidelberg, 2004.

Elangovan K. : *GIS, Fundamentals, Applications and Population Geography*, New India Publishing Agency, New Delhi 2006.

Harvey, David: *Explanation in Geography.* London: Edward Arnold, 1969.

Harvey, Francis : *A Primer of GIS, Fundamental Geographic and Cartographic Concepts,* The Guilford Press, 2008.

Jackson, Peter: *Maps of Meaning: An Introduction to Cultural Geography*. London and Boston: Unwin Hyman, 1989.

James, T.: *The Best Poor Man's Country: A Geographical Study of Early Southeastern Pennsylvania*. Baltimore: Johns Hopkins Press, 1972.

King, Anthony D.: *Buildings and Society: Essays on the Social Development of the Built Enviroment*. London: Routledge & Kegan Paul, 1980.

Lewis, Earl. *In Their Own Interests: Race, Class, and Power in Twentieth-Century Norfolk, Virginia*. Berkeley: University of California Press, 1990.

Lewis, Peirce. *New Orleans: The Making of an Urban Landscape*. Cambridge, MA: Ballinger Publishing Co., 1976. Annotation

Michael P.: *The Making of the American Landscape*. Boston: Unwin Hyman, 1990.

Michael, H.: *City Form and Natural Process: Towards a New Urban Vernacular*. London: Croom Helm, 1984.

Neil, H.: *The Land of Contrasts: 1880-1901,* The American Culture 5. New York: G. Braziller, 1970.

Peter, G.: *The Slow Plague: A Geography of the AIDS Pandemic*. Oxford and Cambridge, MA: Blackwell Publishers, 1993.

Ray Allen: *The Far Western Frontier, 1830-1860*. The New American Nation Series. New York: Harper, 1956.

Richard SCASE : *Managing Creativity,* Open University Press, 2000.

Ritchie, J.: *Landscape and Material Life in Franklin County, Massachusetts, 1770-1860*. Knoxville: University of Tennessee Press, 1991.

Stephen, D.: *Fields of Vision: Landscape Imagery and National Identity in England and the United States*. Cambridge: Polity, 1993.

Victor, A. : *Geography of Art and Culture*, Contributions to Economic Analysis, Elsevier, 2004.

Watkins, Eric: *The Middle Eastern Environment*, London, The British Society for Middle Eastern Studie, 1995.

Wolf, Aaron T.: *Hydropolitics along the Jordan River, Scarce Water and its Impact on the Arab-Israeli Conflict*, NYC, United Nations University Press, 1995.

Index

A

B

C

D

E

G

H

I

L

M

N

O

P

Q

R

S

T

U

W

□□□